URBAN LIFE
IN TRANSITION

Volume 39, URBAN AFFAIRS ANNUAL REVIEWS

URBAN LIFE

IN TRANSITION

Edited by

M. Gottdiener
Chris G. Pickvance

Volume 39, URBAN AFFAIRS ANNUAL REVIEWS

 **SAGE PUBLICATIONS**
The International Professional Publishers
Newbury Park   London   New Delhi

For information address:

SAGE Publications, Inc.
2455 Teller Road
Newbury Park, California 91320

SAGE Publications Ltd.
6 Bonhill Street
London EC2A 4PU
United Kingdom

SAGE Publications India Pvt. Ltd.
M-32 Market
Greater Kailash I
New Delhi 110 048 India

Printed in the United States of America

Library of Congress Cataloging-in-Publication Data

Urban life in transition / edited by M. Gottdiener, C. G. Pickvance.
      p.   cm. — (Urban affairs annual reviews : v. 39)
   Includes bibliographical references.
   ISBN 0-8039-3974-4. — ISBN 0-8039-3975-2 (pbk.)
   1. Urban renewal—United States. 2. Metropolitan areas—Economic aspects—United States. 3. Cities and towns—United States—Effect of technological innovations on. 4. Community development, Urban—United States. 5. Inner cities—United States. 6. Neighborhood—United States. 7. Residential mobility—United States.
   I. Gottdiener, M. II. Pickvance, C. G. III. Series.
HT108.U7 vol. 39
[HT175]
307.76 s—dc20
[307.3'416'0973]   91-25273
                                                CIP

FIRST PRINTING, 1991

Sage Production Editor: Judith L. Hunter

Contents

1

Introduction

M. GOTTDIENER
CHRIS G. PICKVANCE

SINCE THE 1970s urban areas in the United States and in other industrialized nations have been subjected to a series of unprecedented changes. Two of the most fundamental involve, on the one hand, the restructuring of the economic base and the shift from mass industrial production to high technology manufacturing and information processing, and, on the other, the demographic diffusion of population on a massive scale across metropolitan regions. These deep-level transformations have affected, in turn, the very nature of the urban experience. Ways of running the city have shifted considerably. For example, urban politics has moved from the freewheeling days of renewal and generous federal funding through fiscal crisis to the present period of limited resources, privatization of services, and combined public/private partnerships in pursuit of new growth.

Massive shifts in population have virtually cleared U.S. cities of the child-rearing middle class. At the same time new immigrant groups, especially from Asia, Mexico, and the Caribbean, have changed the racial and ethnic mix of our largest cities. As the population has become skewed, growing inequalities have developed between the working poor, the underclass, and the well off, thereby producing an extreme social crisis of inequality and polarization that is reflected in high crime rates, drug problems, and urban terror. Population changes have, in turn, affected local politics and the style of urban political interaction. An entirely new socio-political fabric has been woven in recent years that holds everyday urban life together in ways that differ from the past.

It is remarkable to note both the profound nature of all these changes and the fact that many of the shifts are interrelated despite their diverse causes.

Thus population deconcentration and the suburbanization of people and industry, which is a product of both national government policies and private sector restructuring, has changed in turn the nature of local urban politics and the constitution of city populations. Contrary-wise, local reconstruction as a result of a decade of fiscal crisis and public/private growth partnerships has altered the economic base of urban places, thereby producing, in turn, a different mix of people and employment opportunities than previously characteristic of large cities.

By the 1980s, as a consequence of several decades of change, U.S. cities were pitted in a highly competitive game against each other for economic resources. Cities were abandoned to a Hobbesian fate by limited federal policies. A host of programs were initiated at the local level that advertised themselves as public/private partnerships and whose principal aim was the revitalization of a restructuring urban economy. Social programs over the last two decades have taken a backseat to these more pressing concerns. Given the changed nature of city populations, the retreat from an activist social stance has produced tensions within the city that presently seem only to grow worse.

More recently, signs of an economic turnaround for cities have surfaced. Observers note that restructuring trends favored the city by promoting service industries (Castells, 1986; Noyelle & Stanback, 1984; Sassen, 1989). A renaissance of a kind followed the downturn in manufacturing employment and resulted from growth spurts in producer-related services. Especially pertinent to revival were the complex of activities centered around information processing, real estate, and finance capital. This surge made the city a more specialized node in the global economy with related economic activities, such as manufacturing, carried out elsewhere.

Growth in producer services involved the expansion of the labor force including those professionals commanding relatively high salaries, such as lawyers, administrators, executives, financial analysts, programmers, and the like. This core in turn created the need for new specialized consumer services catering to the upscale tastes of the information economy. Some observers have noted that this complex of restructured economic changes was serviced by a new wave of unskilled immigrants who took up positions in the expanding sector of consumer services (Sassen, 1989). Rounding out this picture new immigrant groups promoted the explosion of services catering to their own specialized consumer needs.

The outcome of metropolitan development in the 1990s, then, is a more specialized urban economy centered principally on the provision of advanced high tech command and control services with a bipolar urban culture based, on the one hand, on the needs of the relatively affluent, upscale professionals and, on the other, on the infusion of new immigrant groups and their culturally distinct service industries. Yet not every city has prospered and restructured quite in this fashion without problems. Places like Houston, for example, encountered a downturn as resource-based industry declined. Detroit never overcame its doughnut-hole development to participate adequately in the new service boom. New York and Boston, which epitomized for some the vitality of restructured economies, were hit hard by finance capital's crisis following the October 1987 events on Wall Street. Uninterrupted expansion has currently turned cyclical. Consequently, new economic woes have been added to decades-long concerns about fiscal well-being and the torn social fabric.

For the most part, academic observers of urban restructuring have focused on the economic aspects of change (see, for example, Beauregard, 1989; Castells, 1986; Harvey, 1987; Kasarda, 1988; Sassen-Koob, 1986; Smith & Feagin, 1987). In some cases, for instance, published accounts will advocate a single perspective that is claimed to explain change in a global way. Much of the "new" urban geography, for example, suggests that it is capitalism itself that has been altered and that has evolved to a new mode called "flexible accumulation" (Boyer, 1986; Harvey, 1987; Leborgne & Lipietz, 1987; Piore & Sabel, 1984; Scott & Storper, 1986). The implications of much of this work suggest a Pollyannaish outcome for places that pursue high-tech seeds of growth and help ease capitalism along in its new birth pangs. There have been substantive challenges to this view recently (Gertler, 1988; Gottdiener, 1990; Peet, 1987; Sayer, 1990).

A second school of thought focuses on deindustrialization and the effect on place of the global capitalist economy, especially the world division of labor (Bluestone & Harrison, 1982; Frobel et al., 1980; Sassen-Koob, 1984; Smith & Feagin, 1987). This approach is also limited and possesses its own critics (Busch, 1989; Gottdiener & Komninos, 1989; Milios, 1989; Peet, 1989; Scott, 1988).

A third approach focuses on technology and proclaims the advent of a new era where advanced producer services and information processing will become the new economic base of the "postindustrial" city (Castells, 1986; Hall & Markusen, 1986; Kasarda, 1988; Noyelle & Stanback, 1984). Postindustrialism, however, has been critiqued for failing to account for demographic and growth changes in the United States (Frey, 1987) and for its exaggerated claims regarding the movement of capitalism to such a new economic base (Gottdiener & Kephart, 1991; Scott, 1988).

In short, the urban transformation involves multicausal forces; diverse changes at economic, political, social, and spatial levels; and, represents a clash of competing arguments that attempt to explain restructuring patterns in economic terms. At present, it seems, no one school of thought dominates, rather they all possess a commonly held understanding that economic changes are the root causes; hence their common concentration on the economic dimension of transformation.

To be sure, the understanding of economic change is central to the study of urban restructuring. But the sum total of such efforts gives the false impression that people and their respective everyday lives do not exist. Consequently, urban economics becomes the privileged field of inquiry. The premise of this volume is quite different. We suggest that the drama of change, its operation at the quotidian level and the impact of restructuring on the neighborhood, the household; on class, race, and gender; and, finally on urban institutions such as political leadership, all require examination in their own right.

The present volume seeks to provide just such a human dimension to the restructuring record. It rejects univocal causes and single factor views on the current state of cities. The chapters in this volume aim for comprehensive treatments of their respective subjects. But beyond that they share a variety of theoretical points of view. What they hold in common is the need to place aspects of social life within the current context of urban restructuring. The aim, then, is to survey the changes in neighborhoods, households, population distribution, and social institutions that are associated with, and in many cases have resulted from, the economic transformations since the 1970s. We all acknowledge the roots of restructuring, but seek to place everyday concerns back on the urban agenda by assessing how urban life has been altered as a consequence of economic change.

This collection contains eleven chapters. The first six papers are more concerned with micro-level effects of change on everyday life including crime, demographic shifts, racism and the underclass, immigration, ethnicity, and gender. The remaining chapters take a more structural look at the impact of economic restructuring on households, privatization and public provision including housing, the role of public/private partnerships and, finally, the role of leadership in urban politics.

George Kephart's leadoff chapter examines the relationship between industrial restructuring and demographic changes in the United States. Much of the literature argues for some direct link between changes in the organization of production and the reorganization of space. Kephart dissents from that position. The effects of reorganization at the level of production have been over-emphasized, such as in the literature on flexible accumulation, just as case studies of high-tech places, such as Silicon Valley, have been overused. Kephart argues for a more loosely coupled and contingently related model of change and demonstrates how economic transformations have had an *uneven* effect on individuals that in turn has altered migration patterns in very selective ways. Most analyses of restructuring ignore the dimension of population change, as if people will automatically seek out those areas that are prospering while forsaking their roots. Kephart's chapter fills the need for a demographic analysis of migration related to aspects of restructuring. He concludes that class effects account for the ability of individuals to adjust to change and, consequently, that a selective sifting of the population along class and racial lines has accompanied restructuring.

Yen Le Espiritu and Ivan Light are interested in the changing nature of ethnicity within cities. Their work is demographically sensitive to the description of ethnic communities in the United States urban system. Thus while New York, Chicago, and San Francisco are well-known for their vibrant cosmopolitan culture, many United States cities have little evidence of ethnic concentrations. In addition, according to the authors, industrial restructuring has had a limited direct effect on ethnic changes. Indirectly, however, shifting patterns of immigration have altered the ethnic mix of population in central cities. The implications of these changes are explored and, most especially, examined as they relate to the shift from the stereotypical Eastern European, Irish, Italian, and Jewish community models to the current importance of Caribbean, Hispanic, Asian, and African influences. The literature on urban ethnicity that has for so long promoted a focus on the former group experience must make way today for new ethnographies depicting the changed ethnic realities of cities.

Robert Bullard and Joe R. Feagin are concerned with racial issues and their urban effects. The increasing segregation within central city areas and the growing concentration of the underclass presents society with its most serious social crisis since the 1960s. Bullard and Feagin argue for the continued importance of racism as a factor in the restructuring of cities that is relatively independent of economic effects. They also join in the controversy regarding the nature of the black family and its relation to the increasingly hopeless condition of the underclass. Their chapter takes issue with the work of Wilson (1987), who suggests that class factors dominate in the creation of the underclass. Instead, they tie the growth of the underclass to the selective effects of global restructuring within a racist society.

Linda McDowell's chapter examines the impact of economic restructuring on women. She outlines trends in the United Kingdom and argues that existing theories are inadequate. She shows that economic restructuring has had various benefits for woman: a rapid growth in female participation in the labor force (simultaneous with a fall in male participation) and greater opportunities for highly qualified women. But the growth in female jobs has been in part-time jobs in the service sector, most of which are female-dominated and give rights to few social benefits, while earnings differentials between men and women remain significant. There is thus both a convergence toward women being working mothers rather than "housewives" and a divergence in earnings levels among women of different occupational levels. At the same time, the family and welfare state have undergone major changes. Far more women live and conceive children outside the traditional family and U.K. divorce rates are the highest ever. Yet the welfare state oscillates between its traditional assumption that women are economically dependent on men and the introduction of antichild and antifamily measures. Its advocacy of "community care" implies a transfer of responsibility for the elderly to women.

McDowell argues that such trends are not explicable in terms of prevailing theories. On the one hand, theories of economic restructuring (such as flexible accumulation theories of Piore and Sabel, 1984, and the "new industrial geography") either ignore its impact on women and neglect the domestic sphere altogether or treat women as marginal workers. On the other hand, socialist feminist theories treat women as a disposable element of a "reserve army of labor" when all the evidence is that they are an essential and permanent part of the labor force. They are also unable to account for the contradictions in welfare policy, and treat women's unpaid domestic labor as a key element in the reproduction of male labor power when in the United Kingdom there is evidence of a reduction in the amount of domestic labor. In conclusion, McDowell argues that theories need to take account both of the divergences among women in their work situation and their common experience due to their responsibility for child care.

Ralph B. Taylor addresses the issue of crime and its effects on urban life. After providing data on the nature, extent, and locations of crime and patterns of change over time, he relates these trends to the record of urban restructuring. Taylor then assesses the different theories of crime and their relative efficacy in explaining new patterns. The discussion ranges across the topics of crime, criminal careers, and, importantly, victimization. Taylor concludes the chapter with an extended discussion of the impact of crime on daily urban life and the adjustments occurring in local neighborhoods to an increasingly risky inner-city experience.

The remaining chapters address the transformations of urban life as they relate to institutional changes. Bryan R. Roberts's work concerns the impact of restructuring on households, especially the less wealthy families living in cities. He compares the cases of Latin America, Britain, and the United States. The privatization of the welfare state in most places has had important effects on families. It has exposed households to market forces and created a greater need for neighborhood networks of self-help. Given increasing economic uncertainty and cutbacks in welfare programs, households have had to alter their coping strategies. As Roberts points out, however, there are no substitutes for enlightened social measures that address the serious issues of poverty, family breakup, and the like. For some time the community question has been taken up by network researchers who have argued that neighboring has declined in favor of networks deployed across space. Roberts suggests that the network approach is misguided because it focuses on individuals rather than households. The latter's coping strategies as a consequence of restructuring resurrects the need for community and highlights the importance of place once again.

For Roberts, coping strategies developed out of particular arrangements in the past that defined the mix between state provision and market self-provision. In some cases, more intense use of exploitable labor occurs with even greater participation in labor markets. In places like Britain, however, where employment opportunities outside the community are quite limited, such strategies can only be differentially applied by the most employable. In other cases, pressure is applied, often by organized labor, for greater state responsiveness to needs produced by restructuring. Rather than a tendency toward greater collective or community social movements, Roberts observes the growing isolation of households at the present time with its consequent ill affects on everyday life.

Ray Forrest focuses on one of the major changes of the last decade, the spread of privatization, and does so within a comparative context by contrasting the U.S. and the U.K. cases. He points out that this term has been used in many ways. The extreme case, where a state-provided financed and regulated service is replaced by a deregulated privately financed and produced service, is rarely encountered. More often privatization refers to shifts to intermediate points between these extremes. The chapter deals both with theories of privatization and the realities of privatization. Theories of state intervention in the 1970s argued that there was a continuous increase in the extent of state intervention in the economy and society of advanced capitalist countries due to "system needs" and political pressures. Forrest shows that the theories of Left and Right both failed to predict the emergence of the partial reversal of state intervention represented by privatization. (It is only partial because state funding and or regulation are often still involved.) He demonstrates that these theories underestimated the negative side effects of state intervention and that "capitalism can neither live with nor live without an expanding sector of goods and services collectively provided."

Forrest then turns to an examination of the sale of public housing in the United Kingdom, which is the largest element in the British privatization program and has been taken as a model for privatization in the United States and elsewhere. He makes a detailed comparison of U.S. and U.K. public housing and emphasizes the different scale, different quality of stock, and different tenant characteristics of each to explain why privatization did not occur in U.S. public housing. Finally he integrates his discussion of theory and empirical evidence by arguing that the prime reason for the popularity among tenants of the privatization of public housing in the United Kingdom is not tenant dissatisfaction with public provision but the bargain prices attached to these dwellings. He points out the paradox that interference with the market to the extent of 70% discounts on market values is needed to produce privatization. This suggests that theories of privatization that emphasize the virtues of the private sector and the defects of the public sector may be operating on too ideological a plane and that the basis of individual support for privatization are far more concrete and economic.

Gregory D. Squires addresses the issue of public/private partnerships. He frames his analysis against the backdrop of what he suggests has been a consistent feature of U.S. urban life, namely, subscription to the ideology of privatism. Privatism involves the state in the support of the private sector in order to encourage the profitability of locating or remaining in the local area. It advocates a basic faith in the market, which local government is expected to support, as the means by which benefits can trickle down to the community as a whole. Hence, privatism in practice does not mean that the market alone is allowed to prevail, but that the market is to be subsidized by local government.

Criticisms of privatism include a failure to recognize the elite biases of markets, the negative effects of market competition between cities, and the historical failure of the policies of privatism to improve the urban quality of life through trickle-down benefits. With the advent of restructuring, furthermore, most public/private partnerships have directed their efforts at nurturing the central city service economy while allowing general employment levels and the quality of urban life to decline. The result has been a progressive racial and class polarization within cities with disastrous social effects.

Squires concludes his provocative chapter with a discussion of some city cases that represent alternatives to the ideology and practice of privatism.

Clarence N. Stone, Marion Orr, and David Imbroscio's chapter is on the reshaping of urban leadership. Complementing quite well Squires's analysis, this chapter explores the reasons for why privatism is so dominant. As the authors suggest, "For more than 40 years the top priority on the urban agenda in America has been the physical development of the city." Stone et al. suggest that subscription to the general ideology of privatism is insufficient by itself to explain the long-standing support to the private sector of local government and its subsequent neglect of the "second city"—the victims of uneven development. Instead they argue for a regime approach to city politics that recognizes the way urban outcomes are less the result of some conspiracy, such as in the view of growth machine theorists, than a consequence of contentious battles fought among various urban constituencies. What we observe when we look at city politics is the result of regime struggle and not some well-ordered, well thought-out program. In this light the place of leadership becomes most critical as a means of rising above the inadequacies of individual regime priorities. Stone et al. therefore conclude their discussion with a focus on the emerging importance of leadership in city politics.

What can be said about the panoramic view offered by the chapters in this volume? One observation is clear. All analyses that privilege the economic dimension stop at the institutional level of society. We will range further in the following pages—down to the street level and into homes stopping just short of gauging the psychic damage resulting from restructuring. The adjustments observed are profound. They affect all aspects of daily life. Clearly too, power—having it or not having it—determines the extent to which change brings costs or benefits. For the most part, the effects of restructuring work involuntarily through various selection processes on the weakest members of society, affecting the quality of neighborhood life and placing new demands on beseiged households.

Ideologies operate to grease the wheels of change as well. Privatization of public services, while significant, also claims much more for its success on political principles than is warranted by actual practice. The failure to apply pure forms of privatization is obscured behind mists of rhetorical appeals from conservative policies that limit social intervention. Ideology masquerading as the political economy of privatism also disguises the real nature of the public/private relationship. We see it exposed as a consistent commitment to government subsidy of the market over the years. But, because cities have survived the onslaught of economic restructuring during the last few decades, it does not follow that they have become better or more attractive places to live. The social quality of life and the economic well-being of place are still very much distanced from each other. These two poles no longer work in tandem as they were once thought to by followers of market supremacy so that locational prosperity does not necessarily bring with it general improvements in life's quality.

Our separate chapters place flesh and blood on the more abstract aspects of urban restructuring. The new urban realities that emerge do not lend themselves easily to analysis by existing approaches that aim for sweeping generalizations of change or that rely on aggregate statistical evaluations to recreate pictures of society. The general gets lost in the particular when we drop down to the street level. Perhaps it's time to rethink our ways of comprehending these new urban realities. The closing chapter, "A Walk Around Town," offers just such a transformed perspective. Future work might require precisely some combination of generalized analysis and ethnographic accounts of everyday life. Clearly, however, we need to pass beyond the abstract categories of class, race, gender; of power and wealth; of institutions and economics, to some greater grasp of what it's like "down there" in our cities, on our streets, and in our homes.

REFERENCES

Beauregard, R. (Ed.). (1989). *Economic restructuring and political response.* Newbury Park, CA: Sage.

Bluestone, B., & Harrison, B. (1982). *The deindustrialization of America: Plant closings, community abandonment and the dismantling of basic industry.* New York: Basic Books.

Boyer, R. (1986). *La flexibilite du travail en Europe.* Paris: La Decouverte.

Busch, K. (1989). Nation-state and European integration: Structural problems in the process of EEC integration. In M. Gottdiener & N. Komninos (Eds.), *Capitalist development and crisis theory: Accumulation, regulation and spatial restructuring* (pp. 123-153). New York: St. Martin's.

Castells, M. (1986). *High technology, space and society.* Newbury Park, CA: Sage.

Frey, W. (1987). Migration and depopulation of the metropolis. *American Sociological Review, 53,* 240-257.

Frobel, F., et al. (1980). *The new international division of labour* (P. Burgess, Trans.). Cambridge, UK: Cambridge University Press.

Gertler, M. (1988). The limits to flexibility. *Transaction Institute of British Geographers, 13,* 419-432.

Gottdiener, M. (1990). Crisis theory and state-financed capital: The new conjuncture in the USA. *International Journal of Urban and Regional Research, 14*(3), 383-404.

Gottdiener, M., & Kephart, G. (1991). The multi-nucleated metropolitan region. In R. King, S. Olin, & M. Poster (Eds.), *Postsuburban California: The transformation of Orange County since World War II.* Los Angeles: University of California Press.

Gottdiener, M., & Komninos, N. (Eds.). (1989). *Capitalist development and crisis theory: Accumulation, regulation and spatial restructuring.* New York: St. Martin's.

Hall, P., & Markusen, A. (1986). *Silicon landscapes.* Boston: Allen & Unwin.

Harvey, D. (1987). Flexible accumulation through urbanization. *Antipode, 19,* 260-286.

Kasarda, J. (1988). Economic restructuring and America's urban dilemma. In M. Dogan & J. Kasarda (Eds.), *The metropolis era: Vol. 1. A world of giant cities* (pp. 56-84). Newbury Park, CA: Sage.

Leborgne, D., & Lipietz, A. (1987). *New technologies, new modes of regulation: Some spatial implications* (Working Paper No. 8726). Paris: CEPREMAP.

Milios, J. (1989). The problem of capitalist development. In M. Gottdiener & N. Komninos (Eds.), *Capitalist development and crisis theory: Accumulation, regulation and spatial restructuring* (pp. 154-173). New York: St. Martin's.

Noyelle, T., & Stanback, T. (1984). *The economic transformation of American cities.* Totowa, NJ: Rowman & Allanheld.

Peet, R. (1987). *International capitalism and industrial restructuring.* Boston: Allen & Unwin.

Peet, R. (1989). Conceptual problems in neo-Marxist industrial geography. *Antipode, 21*(1), 35-50.

Piore, M. J., & Sabel, C. F. (1984). *The second industrial divide: Possibilities for prosperity.* New York: Basic Books.

Sassen, S. (1988). New trends in the socio-spatial organization of the New York City economy. In R. Beauregard (Ed.), *Economic restructuring and political response.* Newbury Park, CA: Sage.

Sassen-Koob, S. (1984). The new international division of labor in global cities. In M. Smith (Ed.), *Cities in transformation.* Beverly Hills, CA: Sage.

Sayer, A. (1990). Post-Fordism in question. *International Journal of Urban and Regional Research, 13*(4), 666-695.

Scott, A. (1988). *Metropolis.* Berkeley: University of California Press.

Scott, A., & Storper, M. (Eds.). (1986). *Production, work, and territory: The geographical anatomy of industrial capitalism.* Boston: Allen & Unwin.

Smith, M., & Feagin, J. (1987). *The capitalist city: Global restructuring and community politics.* Oxford: Basil Blackwell.

Wilson, W. J. (1987). *The truly disadvantaged: The inner city, the underclass, and public policy.* Chicago: University of Chicago Press.

2

Economic Restructuring, Population Redistribution, and Migration in the United States

GEORGE KEPHART

IN THE LAST SEVERAL DECADES, the United States has experienced profound changes in the structure of settlement space. Accelerated population redistribution patterns in the 1970s and 1980s point to a restructuring of settlement space characterized by new and dynamic patterns of uneven development. Formerly dominant metropolitan areas in the Northeast and Great Lakes regions have experienced losses in population and jobs, while newer metropolitan areas themselves have been experiencing extensive deconcentration. The compact central city, surrounded by organizationally dependent residential areas, has become an anachronism. Today's metropolitan areas are a sprawling expanse of spatially diverse activities, governed by a mosaic of local governments. Moreover, non-metropolitan areas, both adjacent and nonadjacent to metropolitan areas, experienced growth rates in the 1970s that exceeded the growth of metropolitan areas; a reversal of previous growth patterns. While the growth rate of nonmetropolitan areas fell slightly below that of metropolitan areas in the early 1980s, nonmetropolitan areas continued to grow at a higher rate than they did in the decades previous to the 1970s.

A central focus of recent research has been on the accelerated patterns of uneven development associated with recent redistribution trends. For example, a growing literature focuses on the inner-city ghettos present in many of the largest metropolitan areas. Many inner-city areas are now experiencing unprecedented poverty, low levels of labor force participation, high crime rates, and social disorganization (Kasarda, 1988; W. J. Wilson, 1987).

Indeed, the level of deterioration in inner-city neighborhoods has led some to suggest the emergence of an economically marginalized and spatially segregated underclass (W. J. Wilson, 1987). Other literature has focused on areas that have benefited from the redistribution of population and jobs; namely, centers of high technology, military production and research, and advanced service provision (Castells, 1985; Noyelle & Stanback, 1984).

A growing literature in sociology, geography, and economics views current patterns of population redistribution and uneven development as principally the result of economic dislocations resulting from the rise of a new spatial organization of production (Bluestone & Harrison, 1982; Castells, 1985; Henderson & Castells, 1987; Noyelle & Stanback, 1984; Scott, 1988a, 1988b). While this literature is broad, the central theme has been that economic restructuring, accelerated by the recessions of the early 1970s and 1980s, has resulted in fundamental changes in both what is produced and how it is produced. It is argued that investment patterns have shifted away from many traditional economic activities in the core regions, such as automobile and steel. Instead, investment has been redirected overseas, to smaller metropolitan and nonmetropolitan areas, and into new types of industries. Furthermore, the diffusion of advanced technologies into the workplace has made possible new organizational, financial, management, and production strategies. These strategies are often spatial. They engender diversified investments, labor sourcing, and the spatial diffusion of production activities across regions and nations.

The purpose of this chapter is to evaluate economic restructuring as an explanation for recent patterns of population redistribution and uneven development by examining the mechanisms through which economic restructuring affects migration. A general research framework for appraising the effect of restructuring on migration will be presented. This framework incorporates the idea that industrial restructuring is only one of a number of contingent societal factors restructuring settlement space (Gottdiener, 1985, 1989). In addition, this framework links societal and individual levels of analysis. It stresses that the effect of economic restructuring on population redistribution will depend on how resultant patterns of local economic restructuring alter the employment experiences of individual workers, and on how the employment experiences of individual workers in turn affect their migration decisions. The uneven effect of local economic restructuring on the employment experiences and migration responses of individuals is important. It is a key mechanism by which economic restructuring is related to patterns of uneven development.

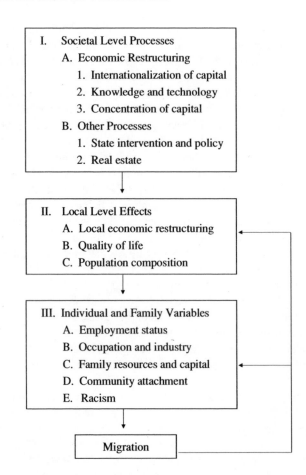

Figure 2.1. A Framework for Examining the Effects of Economic Restructuring on Migration

A FRAMEWORK FOR EXAMINING THE EFFECTS OF INDUSTRIAL RESTRUCTURING ON MIGRATION

Figure 2.1 depicts a general research framework for assessing the effects of economic restructuring on migration. This framework is not intended as a formal model; rather, it is designed to illuminate the relationships among economic restructuring, migration, and population redistribution. Unlike conventional views of economic restructuring, this framework adopts a multilevel approach integrating micro and macro levels of analysis. It suggests that there can be no simple and direct relationship between economic restructuring and population redistribution.

In the top panel of Figure 2.1, economic restructuring is represented as one of several societal processes involved in the social production of space. Economic restructuring includes social changes that are transforming what is produced, how it is produced, and where it is produced (Gottdiener, 1989; Henderson & Castells, 1987; Markusen, 1985; Massey, 1984; Massey & Meegan, 1982; Noyelle, 1987; Peet, 1983; Roberts, Finnegan, & Gallie, 1985; Scott, 1988a, 1988b; Scott & Storper, 1986; Sternlieb & Hughes, 1986; Storper, 1985; Van der Knaap & Wever, 1987; Walton, 1985). Economic restructuring involves several related processes. First, the internationalization of capital and finance is resulting in increased competition for many domestic firms, and fueling the movement of jobs and investments overseas. Multinational corporations are locating production activities to take advantage of cheap overseas labor, advantageous exchange rates, and incentives offered by foreign governments. Second, new technologies have revolutionized production and corporate organization. Technology has resulted in new products and, more importantly, new methods of production. It has made possible the vast spatial dispersion of economic activities, and the increased scales of production and organization characteristic of modern corporations. Finally, economic restructuring involves the emergence of services as a dominant economic sector. There has been a rapid growth in producer services, typified by the growth in advertising, consulting, information processing, and other advanced services in the central business districts of large metropolitan areas. There has, however, also been rapid growth in consumer services. To some degree, low skilled service jobs have offset jobs lost in manufacturing.

It is unlikely that any theory based solely on economic restructuring can account for current redistribution patterns. As indicated by the top panel of Figure 2.1, there are other factors involved in the production of space. Gottdiener (1985, 1989) suggests additional forces that in concert with economic restructuring have played an important role in the restructuring of settlement space since World War II. First, interventions by the state have played a central role in the restructuring of settlement space. For instance, federal banking policies, loan guarantees, subsidies, tax laws, and highway programs have channeled enormous sums of money into the real estate sector (especially single family homes). Moreover, federal, state, and local government policies have figured prominently in urban renewal and economic development projects. Second, a relatively independent circuit of capital, involving real estate, has played a central role in the production of space. As a sector of economic activity, real estate investment, speculation, and development rivals the primary (industrialized) circuit of capital.

The societal processes at the top of Figure 2.1 must affect population redistribution through the pattern and extent of changes they produce in local communities (panel II in Figure 2.1). It is within the context of the local community (place of work and place of residence) that individuals experience the affects of economic restructuring and make migration decisions. Economic restructuring processes will affect local communities primarily by generating changes in the occupational and industrial structure of the local economies (i.e., local economic restructuring). There will be several components of change in the occupational and industrial structure of local economies. These include layoffs and plant closures, the internal restructuring of establishments due to new production techniques and forms of organization, and the growth of new and existing establishments.

Economic restructuring will also affect the quality of life in local communities. Services provided by local governments are a major determinant of quality of life. In turn, the ability of local governments to provide services will depend heavily on the local tax base. Services provided by local governments such as roads, parks, education, and recreation facilities have been shown to play an important role in migration decisions (DeJong & Sell, 1977).

The effect of changes in local economic structures on migration will be mediated by the characteristics of families and individuals, and their role within the local economy (panel III in Figure 2.1). Local economic restructuring will motivate migration primarily through its effect on the employment circumstances of local residents. Local economic restructuring will determine the likelihood that individuals employed in a given occupation and industry, in a given location, experience employment stress (unemployment or underemployment) that may in turn motive them to migrate. Additionally, local economic restructuring will motivate migration via its effect on the structure of local employment opportunities. The extent of local employment opportunities for an individual will depend upon the local restructuring of industries in which they are most likely to seek and find work.[1] Secondarily, local economic restructuring will motivate migration via its effect on the quality of life in local communities.

It is important to recognize that local economic restructuring will unevenly motivate individuals to migrate. The risk of experiencing unemployment and the extent of local employment opportunities will vary depending on the occupation and industry of a worker. Especially in diversified economies, this may result in considerable variation in the effects of local economic restructuring on workers.[2] Similarly, because of the fragmented and segregated structure of local governments within local labor market areas, the effects of restructuring on the quality of life will not be felt equally by workers. The fiscal well-being and resilience of municipalities will vary considerably within metropolitan areas.

It is also important to recognize that motivations to migrate resulting from local economic restructuring will produce uneven migration responses. The individual and family-level factors in panel III of Figure 2.1 will intervene between motivations to move or stay, and migration. This can be conceptualized within a value-expectancy framework (DeJong & Fawcett, 1981). According to this framework, the decision to migrate is a function of a set of values (or goals) and the expectation that a migration will further those goals. Families and individuals vary in terms of their expectations about the various consequences of a move, and in the relative value that they attach to each of these consequences. The recognition that a migration will have a range of consequences for family members is central to this model. A move will affect their proximity to family, friends, social services, and cultural amenities; will alter their life-styles and employment situation; and will have expected costs.

The importance of values and expectancies becomes apparent when the effect of employment circumstances on migration is examined. From the perspective of this framework, persons will display different migration responses to the employment effects of local economic restructuring partly because they will have different expectations about the employment consequences of a move. Differences in expected employment consequences of migration may reflect the expected income gains relative to the cost of moving, the degree to which employment prospects are better in other locations, and the quality of information about employment opportunities in other locations. Differences in expectations about the employment consequences of a move may come from the variations in the potential for employer and industry changes that characterize different careers (Spilerman, 1977). For some workers, changes in employers or even industries can be made with little loss of "tenure." Their skills and experience are transferable to other settings (e.g., engineers). For others, skills and experience will not be transferable across firms and industries (e.g., postal workers). For such workers, unemployment can be devastating since it means the loss of tenure that has been built up over many years, and they are afforded few prospects for comparable employment in other firms.

Based on the value-expectancy model, however, differences in migration responses to the employment effects of local economic restructuring will also result from the relative value placed on employment considerations relative to other factors. The consequences of a move for access to local amenities, crime risks, the quality of schools, social status, and the desire to be near friends and family will be weighed against employment considerations when deciding whether to move.

The mediating role of family and individual factors between local economic restructuring and migration suggests that there may be systematic differences in the migratory responses of families to local economic restructuring. This can have important consequences for both the family and the local community. For example, individuals who do not migrate from areas experiencing economic decline should be less likely to find employment than those who do. For the local community, who moves and who does not will have important consequences for the tax base, which will in turn be associated with the quality of local schools and services. It will also be an important determinant of population composition. This will in turn affect the attractiveness of the area to potential migrants and industries.

In short, *the uneven effect of local economic restructuring on individuals, and individual differences in migratory responses to these effects is an important mechanism by which economic restructuring generates patterns of uneven development.* In areas experiencing economic decline, those who are willing or able to migrate to locations with better economic opportunities will do so. Left behind will be persons who, for whatever reason, are not willing or able to move, or who would not benefit from migration. The tax base will be eroded, service provision will suffer, and the attractiveness of the area to both potential migrants and industry will decline.

In summary, this framework recognizes that economic restructuring is only one of a number of factors restructuring settlement space. Additionally, it specifies the effects of economic restructuring in a manner that is relevant to the migration decision-making process, and it recognizes that the migration response of families to economic restructuring will be uneven. The differential responses of families to local economic restructuring is recognized as an important determinant of the migration patterns generated by economic restructuring.

This framework has two important implications for the explanatory power of economic restructuring processes as explanations of population redistribution. First, the overall effect of economic restructuring processes on population redistribution will depend on the extent to which they generate broad and systematic changes in local economies. *If only a few local econo-*

mies are affected, or if the pattern of effects on local economies is not systematic, then processes of economic restructuring can have little effect on population redistribution. Second, the effect of societal processes of economic restructuring on population redistribution and migration will depend on the extent to which changes generated in local economies drive migration. *If changes in local economies caused by economic restructuring have little effect on migration, then economic restructuring will have little power as an explanation for population redistribution.*

AN EVALUATION OF THE SPATIAL EFFECTS OF ECONOMIC RESTRUCTURING

The literature on industrial restructuring is broad, and my goal is not to conduct a review of this literature. At the risk of oversimplifying, let me suggest that three processes have been emphasized in the economic restructuring literature: (1) deindustrialization, resulting in plant closings and layoffs; (2) the growth and spatial patterns of high technology firms; and (3) the growth in services—especially producer services. Drawing upon the framework in Figure 2.1, as well as some empirical evidence, leads to some important conclusions about the potential effects that each of these processes may have on population redistribution.

DEINDUSTRIALIZATION

One process emphasized in the economic restructuring literature has been *deindustrialization.* A number of researchers (Bluestone & Harrison, 1982; Feagin & Smith, 1987; Massey & Meegan, 1982) have focused on the rash of plant closings and layoffs in labor-intensive manufacturing firms, coinciding with the economic crisis of the 1970s, as an important cause of population redistribution. Plant closing and layoffs in traditional manufacturing activities, such as automobile production, steel, and textiles, have contributed to the economic decline experienced by many older cities.

Deindustrialization is the result of two processes. First, the introduction of new technologies into the production process, such as automation, robotics, and computers, has resulted in capital deepening. Second, deindustrialization has been linked to the rise of a new international division of labor (Feagin & Smith, 1987; Walton, 1985). Plant closings and layoffs in domestic industries are associated with the global labor sourcing strategies used by multinational corporations. Improved communications and transportation technologies have enabled multinational corporations to take advantage of cheap labor in less developed countries (LDCs) and peripheral locations in the more developed countries (MDCs). This has resulted in the growth of manufacturing employment in LDCs and nonmetropolitan areas at the expense of older industrial centers, such as those located in the Northeast and Great Lakes regions.

This perspective contains no explicit theory of the individual migration process, although it can be argued that it makes an implicit assumption that the decision to migrate between labor market areas is largely determined by employment stress. It would imply that job losses in the industrial states have induced migration to areas offering more favorable employment opportunties.

To some degree, current redistribution trends appear to be consistent with deindustrialization. Regions that have traditionally been the nucleus of manufacturing activity in the United States have experienced population loss, while the South, West, and nonmetropolitan areas have experienced growth. Moreover, clear examples of areas that have experienced plant closures followed by dramatic population losses abound. There can be little doubt that deindustrialization has been an important cause of population loss in some areas.

There are reasons to believe that the population redistribution effects of deindustrialization have been exaggerated, however. One of the strongest pieces of evidence to support this conclusion is that unemployment does not play a large role in driving overall levels of migration. Kephart (1990), using longitudinal data from the Panel Study of Income Dynamics, found that the vast majority of intercounty migrations were not preceded by unemployment. Moreover, unemployed workers are actually *less* likely to migrate between counties than employed workers, especially if living in a county with a high unemployment rate.[3]

Data from the Annual Housing Surveys (AHS) provide additional information on the extent to which migrations are driven by employment-related factors.[4] The AHS, a nationally representative survey of housing units collected by the U. S. Bureau of the Census, provides a rich source of data on reasons for migration. L. Long (1988, p. 235) has tabulated reasons for interstate moves based on several years of the AHS data. Table 2.1 presents the distributions of reasons for interstate moves made in the 12 months preceding the 1979, 1980, and 1981 surveys. It can be seen that job transfers, looking for work, and taking a new job account for about 47% of the reasons for moving. However, 22.2% of employment-related moves were job transfers, and thus are unlikely to be associated with employment stress. This is

TABLE 2.1

Selected Main Reasons [a] for Household "Reference Persons" Moving
Between States in the 12 Months Preceding the 1979, 1980, and 1981 Annual Housing
Surveys (in thousands)

	Percentage
Job transfer	22.2
To look for work	6.3
To take a new job	18.7
Entered or left armed forces	3.4
Retirement	2.4
Attend school	5.6
Other employment reasons	3.2
Divorced or separated	2.6
To be closer to relatives	8.6
Other family reasons	3.0
Wanted change of climate	6.0
Other reasons	15.8
Not reported	2.0
Total (n = 6,250)	100.0

a. All categories accounting for at least 2% of interstate moves made by household reference persons.
SOURCE: Reprinted from *Migration and Residential Mobility in the United States*, Table 7.1, by Larry Long, © 1988 Russell Sage Foundation. Used by permission of the Russell Sage Foundation.

probably also true of many of those taking new jobs as well. Only 6.3% of all moves were made to look for work.

Other evidence suggesting the limited role of plant closings and layoffs in driving migration is to be found in aggregate rates of migration. Rustbelt states such as Ohio, Pennsylvania, Michigan, and New York actually experienced lower rates of out-migration than other states in the 1970s (L. Long, 1988). The population losses experienced by these states were the result of even lower rates of in-migration. This is inconsistent with the view that population losses in these states were due to high rates of out-migration resulting from plant closures and layoffs.

The research framework presented above suggests that one of the most important consequences of deindustrialization will result not from the volume of migration it generates, but from the way it selects migrants. In areas exposed to plant closures and layoffs, workers who are willing and able to improve their employment chances through migration do so. Left behind are those who, for whatever reasons, do not migrate. Those who do not migrate are likely to become increasingly marginalized, experiencing frequent spells of unemployment interspersed with low-quality jobs (Harrison, 1982; Kephart, 1990; Sandefur, Tuma, & Kephart, in press).

Exploring the consequences of migration response to deindustrialization for individuals and communities may prove useful for understanding linkages between economic restructuring, uneven development, and class. Why the unemployed do not move, and how this is related to the changing structure of the economy is an important direction for future research.

THE GROWTH OF HIGH TECHNOLOGY INDUSTRIES

A second process emphasized in the economic restructuring literature is the rise of new types of industrial activity—especially high technology firms (Castells, 1985; Markusen, 1985; Scott, 1988a). A large number of studies have focused on the growth of technopoles and R & D centers in places such as Orange County and the Silicon Valley in California (Soja, Morales, & Wolff, 1983; Scott, 1988a, 1988b). This literature has noted the rapid growth that high technology firms experienced while traditional manufacturing industries were experiencing decline.

The spatial organization of high technology firms has been emphasized. High technology industries, which are closely coupled with defense spending, have sustained the economy of a number of locations outside traditional cities. Scott (1988a, 1988b) and others have argued that high technology firms exhibit new types of agglomeration tendencies, and that they are the product of a new stage of capitalism in which technology has become a principal source of socio-spatial change.

High technology firms may display unique forms of spatial organization; however, their potential role in generating recent redistribution trends is limited. High technology firms employ a relatively small share of the U.S. labor force, and are thus unlikely to be a major source of regional employment shifts. Of course, the percentage of all U.S. workers employed in high technology firms is dependent on how high technology firms are defined. While the most liberal definition places high technology employment at 16% of the labor force, most definitions place the figure at about 5% (Thompson, 1988).

Additionally, Gottdiener and Kephart (in press) found that high technology manufacturing was not a consistent source of employment growth in deconcentrated counties adjacent to the largest metropolitan areas. They compared the sources of 1975-1980 employment growth among 22 counties that were similar to Orange County, California, in the sense that they were large, urbanized, deconcentrated counties located adjacent to metropolitan areas of more than 1 million population. All had large residential and working populations, were growing, and with one exception (Silicon Valley) contained no large central cities. Using County Business Patterns data (U.S. Bureau of the Census, 1984), employment growth shares were calculated for high technology manufacturing, other manufacturing, services, and several other industrial sectors. In all but one of the counties, high technology manufacturing's share of employment growth was less than that of other manufacturing firms. Moreover, while high technology was a major source of employment growth in a few of the counties, it comprised a small source of employment growth in many others.

In summary, while high technology firms undoubtedly play a prominent role in the spatial restructuring in some locations, they comprise a fairly small component of the employment base in most local economies. As a result, their role in driving population redistribution must be limited. The biggest effect of high technology is likely to operate through its effect on the production process in firms generally, rather than through the spatial dynamics of high technology firms.

THE GROWTH IN SERVICES

A third process emphasized in the economic restructuring literature is the growth of advanced services. It is argued that an economic base dominated by manufacturing is being replaced by one dominated by information- and knowledge-based industries (Castells, 1985; Noyelle & Stanback, 1984). This process results in changes in spatial structure for several reasons. First, the locational patterns of services are different than the locational patterns of manufacturing. Thus compositional effects will result in growth of areas specialized in services, and decline in areas specialized in most types of manufacturing. Second, in knowledge-based industries, production is less tied to particular locations for both the firm and the worker. Finally, services exhibit somewhat unique locational and agglomeration tendencies.

One of the most prominent views of the spatial consequences of the manufacturing to service shift is provided by Noyelle and Stanback (1984). According to Noyelle and Stanback, corporate headquarters, banking, and advanced services have agglomeration tendencies resulting from the necessity of complex linkages and contractual arrangements between firms. This is evidenced, they argue, by the concentration of these activities within the nation's largest metropolitan areas. In conjunction with changes in the organization of production that are producing smaller shares of jobs in production establishments and consumer services, and greater shares in high-level services, the agglomeration tendencies of high-level services within large metropolitan areas form the basis for their sustained dominance and growth. Thus, they contend, many metropolitan areas will ultimately prosper due to their important and strong functional role in the metropolitan hierarchy. In the Northeast and Great Lakes regions, diversified metropolitan areas will experience a period of net out-migration because manufacturing declines will offset gains from new activities as these cities make the transition to advanced service and control centers; however, this decline will be a temporary transitional state. In the long run, according to this perspective, the largest metropolitan areas are accorded the strongest growth prospects.

Noyelle and Stanback's (1984) thesis that changes in the organization of production imply polarized growth tendencies, with the greatest growth accruing to large metropolitan areas, has its foundation in two suppositions. The first is that the economic restructuring is resulting in a transformation of the industrial and occupational structure away from low-level services and manufacturing to high-level services. They point out that the consumer services have not, as is commonly assumed, been the main source of service growth; rather, most of the growth has been in high-level services ("producer services"), including finance, insurance, and real estate; central administrative and auxiliary establishments; business services; and professional services. Assuming that this trend continues, they anticipate that high-level services will account for the largest share of national employment growth. The second supposition is that high-level services have concentrated and will continue to concentrate within diversified metropolitan centers.

Noyelle and Stanback's thesis has received considerable attention, and has even been used to portray the redistribution trends predicted by theories of economic restructuring generally (see Frey, 1987, 1989; Frey & Speare, 1988). Noyelle and Stanback's conclusions about the redistribution consequences of economic restructuring, however, are not shared by much of the restructuring literature. While many restructuring theorists would agree that the largest metropolitan areas will continue to play an important *functional* role as the location of advanced services and corporate headquarters, it does not necessarily follow that this will result in *population concentration.* Furthermore, many have suggested that restructuring will result in the deconcentration of advanced services as well as routine manufacturing activities (Castells, 1985; Hirschhorn, 1985; Holmes, 1986; Mills, 1986; Scott & Storper, 1986). Indeed, even Noyelle and Stanback acknowledge that the evidence on the agglomeration tendencies of high-level services is mixed.

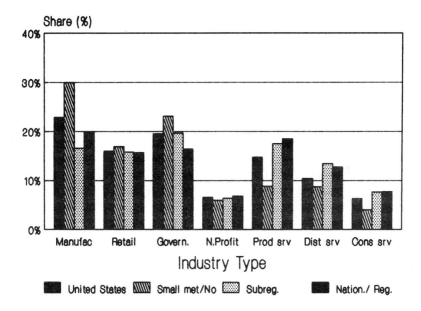

Figure 2.2. Employment Shares of Selected Industry Groups by SMSA Type, 1976
SOURCE: Adapted from Noyelle and Stanback (1984), Table 4.12

For example, they found evidence for the deconcentration of some high-level services such as insurance.

The empirical evidence that Noyelle and Stanback (1984) present on the changing distribution of employment by industry provides weak support for their arguments. Figures 2.2 and 2.3 are adapted from some of the analysis presented by Noyelle and Stanback.[6] Figure 2.2 displays the employment composition, by industry, for small metropolitan and nonmetropolitan areas, subregional diversified service centers, and national and regional diversified service centers. Figure 2.3 displays the contribution of each industrial group to the total employment growth within each metropolitan classification. It is clear from these figures that diversified national, regional, and subregional service centers were more concentrated in producer services than were small metropolitan and nonmetropolitan areas. However, all three metropolitan groups display a high degree of diversity in their economies. Small metropolitan and nonmetropolitan areas contain high percentages of employment in retail and government. In diversified service centers, producer services ac-

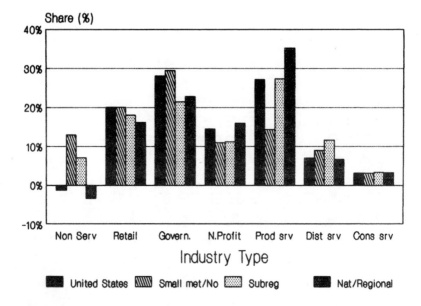

Figure 2.3. Employment Growth Shares of Selected Industry Groups by SMSA Type, 1969–1976

SOURCE: Adapted from Noyelle and Stanback (1984), Table 4.8

count for less than 20% of total employment. It is not surprising, however, that producer services are concentrated in diversified service centers given that Noyelle and Stanback's classification of metropolitan areas is based on their industrial structure.

In line with Noyelle and Stanback's argument, it is also evident from Figure 2.3 that industry growth patterns within metropolitan classifications appear to favor the concentration of producer services in diversified service centers, and to a limited extent, the concentration of nonservice employment in small metropolitan and nonmetropolitan areas. It is interesting to note, however, that in small metropolitan and nonmetropolitan areas government, retail, and producer services all accounted for higher shares of employment growth than did nonservice employment. Smaller areas are also making a manufacturing to services transition, and high-level services are an important component of their economic growth.

However, Noyelle and Stanback's (1984) analysis does not support their assertions about concentration of high-level services at the top of the metro-

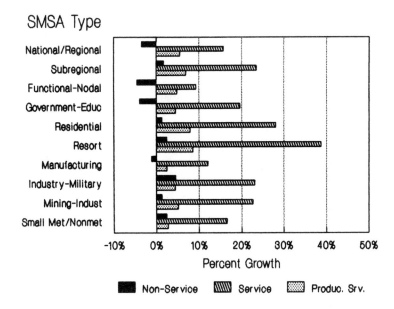

Figure 2.4. Percentage Growth of Industry Groups by Type of SMSA, 1969–1976

SOURCE: Adapted from Noyelle and Stanback (1984), Table 4.8

politan hierarchy. While the data in Figures 2.2 and 2.3 tell us about the composition and change of industry structure *within* standard metropolitan statistical area (SMSA) classifications, they are inappropriate for assessing changes in industrial structure *across* the metropolitan hierarchy. Noyelle and Stanback's analysis examines patterns of concentration within SMSA classifications, while ignoring the differential employment growth between SMSA classifications. The problem can be illustrated as follows. If an industry accounts for 10% of employment growth in area A and 20% of the growth in area B, this does not imply that industry is growing faster in area B than in area A. If area A is growing at 100% and area B at 25%, then the industry would be growing at 10% in area A and only 5% in area B.

It is possible to compute absolute levels of employment growth by industry and SMSA type from Noyelle and Stanback's analysis (Table 4.8, pp. 78-79). Growth in employment for nonservice, service, and producer services are displayed in Figure 2.4. When presented in this fashion, the data fail to support Noyelle and Stanback's conclusions. Evidence for the polarization of manufacturing and high-level services between the top and the bottom of the metropolitan hierarchy is weak. Service growth in small metropolitan and nonmetropolitan areas is larger than in national and regional diversified service centers. Indeed, the greatest growth in both overall service employment and producer services is in metropolitan areas specialized as residential and resort centers. Likewise, there is little evidence that manufacturing is shifting to the bottom of the metropolitan hierarchy.

While Noyelle and Stanback (1984) provide additional data to buttress their claims of agglomeration tendencies of high-level services within diversified service centers, other literature often presents a different picture. Only certain types of advanced services and headquarters appear to locate in the largest metropolitan areas, and it is not clear that those that have necessarily did so to take advantage of agglomeration economies. A substantial literature suggests the deconcentration of advanced services, headquarters, and white-collar work (Armstrong, 1972; Hirschhorn, 1985; Holmes, 1986; Malecki, 1986; Stanback, 1985). Evidence suggests that while financial institutions and the largest corporate headquarters do tend to be located in the central business districts (CBDs) of the nation's largest metropolitan areas, there is still a strong tendency to decentralize coordinating service functions, research and development, and lower level services.

Additionally, some evidence suggests that the growth of downtown office buildings that has occurred in the last several decades may have more to do with urban renewal programs and economic development efforts than with agglomeration tendencies (Anderson, 1964; Armstrong, 1972; Friedland, 1983; Gottdiener, 1985, 1989). In particular, banks and financial institutions realized large profits from the financing of urban renewal efforts, combined with the speculative building of large office buildings in downtown areas. In short, the boom in office growth in CBDs may be largely due to real estate speculation by financial firms, operating in concert with local officials.

In summary, the manufacturing to services shift may be an important mechanism by which economic restructuring is affecting population redistribution. The growth in service industries has been widespread and massive, but the spatial consequences of this shift are still unclear. Mapping the spatial patterns of the growth in services is an important direction for future research.

HUMAN ECOLOGICAL THEORY
VERSUS ECONOMIC RESTRUCTURING

In a number of recent publications, William H. Frey (1987, 1989; Frey & Speare, 1988) has portrayed human ecological theory and economic restructuring as competing theoretical perspectives on the causes of recent redistribution trends. Frey names these perspectives the "deconcentration" perspective and the "regional restructuring" perspective.

Frey's deconcentration perspective is drawn primarily from the human ecological literature (J. F. Long, 1981; Wardwell, 1977, 1980; F. D. Wilson, 1984). Its focus is on locational preferences of individuals and firms, combined with the reduced institutional, monetary, and technological constraints on realizing these preferences. People in the United States, it is argued, prefer lower density locations with ample amenities and good climates. New technologies (especially in the area of transportation and communication) and increased personal wealth have allowed these locational preferences to play an increasing role in the migration decision-making process. Furthermore, reductions in differential factor costs, marketing opportunities, and service provision across space (i.e., past deconcentration) provide an additional impetus for deconcentration, since they facilitate migration to areas that were once remote.

Frey's regional restructuring perspective is drawn almost exclusively from the work of Noyelle and Stanback (1984), although he suggests that it characterizes much of the literature on economic restructuring. The regional restructuring perspective asserts that transformations in the organization of production will result in declines in manufacturing employment in older and larger metropolitan areas. Manufacturing jobs will be relocated to nonmetropolitan areas, overseas, and to smaller metropolitan areas located in the South and West. At the same time, however, many large metropolitan areas will gain new jobs and functions due to their expanding specialization in corporate headquarters, banking, and advanced services. This will occur especially in diversified national, regional, and subregional service centers. Thus restructuring will produce polarized growth tendencies between the top and the bottom of the metropolitan hierarchy.

The research framework presented above suggests that Frey's characterization of the literature is overly simplistic and artificial. Literature on the migration decision-making process (DeJong & Gardner, 1981) clearly shows that individuals display a diversity of migration behaviors within a diversity of community contexts. The motivations for migration cannot be reduced to a simple desire for less densely settled locations, nor are local amenities independent of local economic change. Furthermore, the restructuring literature cannot be characterized solely by the work of Noyelle and Stanback (1984).

Most important, however, Frey trivializes the linkages between macro- and micro-level processes. The deconcentration perspective is primarily a micro-level approach. It accounts for deconcentration in terms of the individual and family decision to migrate. The perspective slights macro-level factors, placing them in an amorphous residual category of "reduced constraints on location." On the other hand, the regional restructuring perspective is a macro-level approach that accounts for regional change in terms of changes in the organization of production. This perspective contains no explicit theory of the individual migration process, although it can be argued that it makes an implicit assumption that the decision to migrate between labor market areas is largely determined by job- and wage-related factors.

A more suitable formulation would recognize that families and individuals migrate for a variety of reasons, including both job and quality of life factors. Macro-level factors, of which economic restructuring is only one, are important insofar as they structure the immediate context in which individuals and families make migration decisions. Viewed in this light, the regional restructuring perspective and the deconcentration perspective are not contradictory, but represent views of the migration process at different levels of analysis. We need to integrate the two perspectives so as to better understand the role of economic restructuring in motivating migration, constraining migration, and structuring the economic context in which migration decisions are made.

CONCLUSIONS

The purpose of this chapter has not been to refute theories of economic restructuring; rather, the goal has been to situate economic restructuring within a broader framework linking societal-level processes with migration, population redistribution, and uneven development. This framework recognizes that economic restructuring is only one of a number of contingent societal processes associated with the social production of space. Moreover, the framework stresses that an adequate theory of the effect of economic restructuring of migration must consider both macro- and micro-level processes. The literature on economic restructuring, while positing that economic restructuring processes are a dominant determinant of population redistribution, has failed to specify the mechanisms by which economic restructuring drives migration at the individual level.

Economic restructuring, as well as other macro-level processes, is relevant to the migration process insofar as it structures the economic and social context within which families make migration decisions. Economic restructuring drives migration and population redistribution through the patterns of change it generates in the economic structure and quality of life of local communities. Local economic changes, in turn, differentially motivate the migration of local residents depending on their occupation and industry of employment, and community of residence. The framework also recognizes that families will display differential migration responses to pressures brought on them by local economic restructuring. Job-related considerations are not the only factors that are associated with migration at the micro level. Families will place different weights on the importance of job-related factors relative to considerations such as their attachment to community, the value of local services, and proximity of friends and family.

Two important directions for future research are suggested by this framework. First, the patterns of effects of societal processes of economic restructuring on local economies needs to be mapped out. To date, too much of the research on economic restructuring has focused on special cases, such as high-technology growth centers, corporate centers, and areas experiencing plant closures. While understanding the spatial strategies employed by different types of industries is important, we must also understand the prevalence and aggregate implications of these strategies.

Second, additional work needs to address the effects of local economic change on migration. At present, our knowledge of the effect of local economic restructuring on the labor force participation and migration of local residents is limited. What we do know is based primarily on aggregate data. An investigation of these relationships is likely to reveal important linkages between economic restructuring and patterns of uneven development.

NOTES

1. Local employment opportunities can have both a relative and an absolute interpretation. Generally, local employment opportunities need to be considered relative to other areas to which the worker might consider moving.

2. Local economies are often characterized by a high degree of interdependence between firms. Accordingly, changes in one economic sector (especially core sectors) will in time be felt by many other economic sectors. As a result, the employment consequences of local economic changes on workers may be diverse at one point in time, but are likely to be more highly correlated over time.

3. The lower rates of out-migration of the unemployed in high unemployment rate counties relative to those in low unemployment rate counties is probably the result of the selection process. As an area experiences economic decline, those who are willing or able to migrate do so. Those remaining thus represent an increasingly immobile group.

4. After 1981, the Annual Housing Survey was renamed the American Housing Survey when it became biannual.

5. Of course, it is possible that high technology firms may generate substantial multiplier effects. This has not, however, been substantiated empirically (See Gottdiener & Kephart, in press).

6. Producer services include finance, insurance, and real estate; central administrative and auxiliary establishments; business services, legal services, and miscellaneous professional services. Distributive services include wholesale, and transportation and communication. Consumer services include personal services, hotel and lodging, miscellaneous repair services, recreation, and private households (Noyelle & Stanback, 1984, pp. 8-10).

REFERENCES

Anderson, M. (1964). *The federal bulldozer*. Cambridge, MA: MIT Press.

Armstrong, R. (1972). *The office industry*. Cambridge, MA: MIT Press.

Bluestone, B., & Harrison, B. (1982). *The deindustrialization of America: Plant closings, community abandonment and the dismantling of basic industry*. New York: Basic Books.

Castells, M. (Ed). (1985). *High technology, space and society*. Beverly Hills, CA: Sage.

DeJong, G. F., & Fawcett, J. T. (1981). Motivations for migration: An assessment and a value-expectancy research model. In G. F. DeJong & R. W. Gardner (Eds.), *Migration decision making: Multidisciplinary approaches to microlevel studies in developed and developing countries* (pp. 13-58). Elmsford, NY: Pergamon.

DeJong, G. F., & Gardner, R. W. (Eds.). (1981). *Migration decision making: Multidisciplinary approaches to microlevel studies in developed and developing countries*, Elmsford, NY: Pergamon.

DeJong, G. F., & Sell, R. (1977). Population redistribution, migration and residential preferences. *Annals of the American Academy of Political and Social Sciences, 429*, 130-144.

Feagin, J., & Smith, M. (1987). Cities and the new international division of labor. In M. Smith & J. Feagin (Eds.), *The capitalist city: Global restructuring and community politics*. Oxford, UK: Basil Blackwell.

Frey, W. H. (1987). Migration and depopulation of the metropolis: Regional restructuring or rural renaissance? *American Sociological Review, 52*, 240-257.

Frey, W. H. (1989). Migration and metropolitan decline in developed countries. *Population and Development Review, 14*(4), 595-628.

Frey, W. H., & Speare, A. (1988). *Regional and metropolitan growth and decline in the United States*. New York: Russell Sage.

Friedland, R. (1983). *Power and crisis in the city: Corporations, unions and urban policy*. New York: Schocken.

Gottdiener, M. (1985). *The social production of urban space*. Austin: University of Texas Press.

Gottdiener, M. (1989). Crisis theory and socio-spatial restructuring: The U.S. case. In M. Gottdiener & N. Komninos (Eds.), *Capitalist development and crisis theory: Accumulation, regulation and spatial restructuring*. New York: St. Martin's.

Gottdiener, M., & G. Kephart. (in press). The multinucleated metropolitan region: A comparative analysis. In R. King, S. Olin, & M. Poster (Eds.), *Post suburban California: The transformation of Orange County since World War II*. Los Angeles: University of California Press.

Harrison, B. (1982). *Rationalization, restructuring, and industrial reorganization in declining regions*. Cambridge, MA: Joint Center for Urban Studies of MIT and Harvard University.

Henderson, J., & Castells, M. (Eds.). (1987). *Global restructuring and territorial development*. Beverly Hills, CA: Sage.

Hirschhorn, L. (1985): Information technology and the new services game. In M. Castells (Ed.), *High technology, space and society*. Beverly Hills, CA: Sage.

Holmes, J. (1986). The organization and locational structure of production subcontracting. In A. J. Scott & M. Storper (Eds.), *Production, work and territory: The geographical anatomy of industrial capitalism*. Boston: Allen & Unwin.

Kasarda, J. D. (1988). Jobs, migration and emerging urban mismatches. In M. G. H. McGeary & L. E. Lynn, Jr. (Eds.), *Urban Change and Poverty* (pp. 148-198). Washington, DC: National Academy Press.

Kephart, G. (1990, May). *The effect of county unemployment rates on out-migration from United States counties: A reappraisal*. Paper presented at the annual meetings of the Population Association of America, Toronto.

Long, J. F. (1981). *Population deconcentration in the United States* (Special Demographic Analysis No. CDS-81-5). Washington, DC: Government Printing Office.

Long, L. (1988). *Migration and residential mobility in the United States*. New York: Russell Sage.

Malecki, E. J. (1986), Technological imperatives and modern corporate strategy. In A. J. Scott & M. Storper (Eds.), *Production, work and territory: The geographical anatomy of industrial capitalism*. Boston: Allen & Unwin.

Markusen, A. (1985). *Profit cycles, oligopoly and regional development*. Cambridge, MA: MIT Press.

Massey, D. (1984). *Spatial divisions of labor: Social structure and the geography of production*. London: Macmillan.

Massey, D., & Meegan, R. (1982). *The anatomy of job loss*. London: Methuen.

Mills, E. S. (1986). Service sector suburbanization. In G. Sternlieb & J. W. Hughes (Eds.), *America's new market geography*. New Brunswick, NJ: Rutgers University Press.

Noyelle, T. J. (1987). *Beyond industrial dualism: Market and job segmentation in the new economy*. Boulder, CO: Westview.

Noyelle, T. J., & Stanback, T. M., Jr. (1984). *The economic transformation of American cities*. Totowa, NJ: Rowman & Allenheld.

Peet, R. (1983). Relations of production and the relocation of United States manufacturing industry since 1960. *Economic Geography, 59*(2), 112-143.

Roberts, B., Finnegan, R., & Gallie, P. (Eds.). (1985). *New approaches to economic restructuring: Unemployment and the social division of labor*. Manchester, UK: Manchester University Press.

Sandefur, G. D., Tuma, N. B., & Kephart, G. (in press). Race, local labor markets, and migration in the United States, 1975-1983. In J. Stillwell & P. Congdon (Eds.), *Modelling migration: Macro and micro perspectives*. London: Bellhaven.

Scott, A. J. (1988a). *Metropolis: From the division of labor to urban form*. Berkeley: University of California Press.

Scott, A. J. (1988b). *New industrial spaces.* London: Pion Limited.

Scott, A. J., & Storper, M. (Eds.). (1986). *Production, work and territory: The geographical anatomy of industrial capitalism.* Boston: Allen & Unwin.

Soja, E., Morales, R., & Wolff, G. (1983). Urban restructuring: An analysis of social and spatial change in Los Angeles. *Economic Geography, 59*(2), 195-230.

Spilerman, S. (1977). Careers, labor market structure, and socioeconomic achievement. *American Journal of Sociology, 83,*551-593.

Stanback, T. M. (1985). The changing fortunes of metropolitan economies. In M. Castells (Ed.), *High technology, space and society.* Beverly Hills, CA: Sage.

Sternlieb, G., & Hughes, J. W. (Eds.). (1986). *America's new market geography: Nation, region and metropolis.* New Brunswick, NJ: Rutgers University Press.

Storper, M. (1985). Technology and spatial production relations: Disequilibrium, interindustry relationships, and industrial development. In M. Castells (Ed.), *High technology, space and society.* Beverly Hills, CA: Sage.

Thompson, J. (1988). Some problems with R&D/SE&Y-based definitions of high technology industry. *Area, 20*(3), 265-277.

U.S. Bureau of the Census. (1984). *County business patterns, 1974-1984.* (Computer file). Ann Arbor, MI: Inter-University Consortium for Political and Social Research (distributor).

Van der Knaap, G. A., & Wever, E. (Eds.). (1987). *New technology and regional development.* London: Croom Helm.

Walton, J. (1985). *Capital and labor in the urbanized world.* London: Sage.

Wardwell, J. (1977). Equilibrium and change in non-metropolitan growth. *Rural Sociology, 42,*156-179.

Wardwell, J. (1980). Toward a theory of urban-rural migration in the developed world. In D. Brown & J. Wardwell (Eds.), *New directions in urban-rural migration* (pp. 71-118). New York: Academic Press.

Wilson, F. D. (1984). Urban ecology, urbanization and systems of cities. *Annual Review of Sociology, 10,*283-307.

Wilson, W. J. (1987). *The truly disadvantaged: The inner city, the underclass and public policy.* Chicago: University of Chicago Press.

3

The Changing Ethnic Shape
of Contemporary Urban America

YEN LE ESPIRITU
IVAN LIGHT

FOCUSING ON MACRO-ECONOMIC CHANGES, much of the literature on restructuring neglects to investigate the ramifications that these changes have on urban everyday life (Piore & Sabel, 1984; Smith & Feagin, 1987). In particular, it ignores the links between macro-economic changes and the changing population mix within American cities. Because global restructuring dislocates people in their homelands and creates a demand for them in other countries, it leads to massive population shifts across national boundaries (Portes & Walton, 1981; Sassen, 1988). As Soja, Morales, and Wolff (1983, p. 219) have suggested, the large and growing streams of immigrants into American cities are tied directly into the restructuring process. As new immigrants enter the city, they change the ethnic shape of contemporary America, altering both the role and status of American ethnics.

Since the 1970s, immigration to the United States has accelerated to levels not seen since the peak years of immigration before World War I. From 1968 through 1990, about 10 million people immigrated to the United States: 4 million came in the 1970s, and another 6 million entered in the 1980s (Dinnerstein, Nichols, & Reimers, 1990, p. 266). The geographic origins of the new immigrants differed from the old. Pre-World War I immigrants were overwhelmingly European. Between 1900 and 1965, 75% of all immigrants to the United States were from Europe. In contrast, today's immigrants come largely from Third World countries. Since 1968, more than 75% of the immigrants traced their ancestry to Latin America and Asia (Dinnerstein et al., 1990; Massey, 1981). Of the immigrants arriving in California during the 1970s, one-third came from Asian countries, and fully 43% came from a single country, Mexico (Muller & Espenshade, 1985). Socioeconomically, today's immigrants are much more heterogeneous than in earlier times. While the vast majority of immigrants before 1920 were unskilled workers, the new immigrants cluster bimodally in both blue- and white-collar occupations (Massey, 1981, p. 59). Commenting on the diversity of contemporary immigration, Portes and Rumbaut (1990, p. 7) wrote, "Never before has the United States received immigrants from so many different countries, [and] from such different social and economic backgrounds."

As in the past, contemporary immigrants flow overwhelmingly toward American cities. Between 1960 and 1979, a central city was the destination of about 55% of the immigrants who were admitted to this country. Towns were the clear second choices of these immigrants. Rural areas were a distant last (Briggs, 1985, pp. 155-156). In 1987, less than 7% of the eight largest immigrant groups chose nonurban areas as their destination (Portes & Rumbaut, 1990, p. 41). However, although an urban phenomenon, immigration does not affect all cities equally. In 1987, immigrant settlement in San Francisco, Los Angeles, Chicago, New York, Washington, D.C., and Miami accounted for 42% of that year's legal immigration (Portes & Rumbaut, 1990, p. 37). The new population mix has produced new urban diversities. For example, Dominican, Haitian, Colombian, and El Salvadoran newcomers (along with Puerto Ricans) add a distinct Caribbean and Latin flavor to New York City, which has been known for its Asian, black, Irish, Jewish, and Italian ethos. Similarly, the city of Los Angeles has become a Third World metropolis with the arrival of immigrants from previously underrepresented places like Soviet Armenia, Iran, and Vietnam (Dinnerstein et al., 1990; Soja et al., 1983).

Theories coined to explain the experiences of European immigrants are not very helpful in understanding the new ethnic realities. The ethnic and socioeconomic backgrounds of the new immigrants are too diverse to fit neatly within the standard assimilation framework. The changing ethnic configuration within American cities raises new questions regarding intergroup relations, immigrant adaptation, and the formation and definition of ethnic communities. This chapter reviews recent research on the new population mix within cities and discusses new approaches to the study of urban ethnicity.

INTERGROUP RELATIONS

Sociologists continue to debate the etiology of intergroup conflicts. Most of the dialogue has been structured around a confrontation between race-based and class-based theorists. According to race-based theorists, prejudice—unfavorable attitudes toward a racial group—causes intergroup conflicts. Prejudice is often attributed to preadult socialization: Children and adolescents acquire prejudice along with other attitudes and values that are normative in their environments (Allport, 1958; Middleton, 1976). In contrast, class-based theorists insist that competition for scarce resources leads to racial antagonism (Bonacich, 1972; Cummings, 1980). An example of the class-based approach is the split labor market theory. Split labor markets arise when there is a large differential in price of labor for the same occupation. Because employers seek the cheapest labor, it is in the interest of high-priced labor to seek to exclude low-priced labor. When the labor market is split along racial lines, economic competition between high-priced and cheap labor is transformed into racial antagonisms (Bonacich, 1972).

Whatever their theoretical interest, students of intergroup relations have primarily focused on racial conflicts between blacks and whites (Blauner, 1969; Bonacich, 1976; Kinder & Sears, 1981). In so doing, they have largely ignored antagonisms among nonwhite ethnic minorities. Conflict among nonwhites is distinct from traditional black-white antagonisms. In black-white conflicts, the two groups involved are distinguishable in terms of the power resources they possess, whites having more of every resource; in contrast, conflicts among nonwhites occur principally between disadvantaged groups (Oliver & Johnson, 1984, p. 75). In light of the substantial influx of immigrants to the United States from Latin American and Asia, researchers need to be more sensitive and concerned with ethnic antagonisms among nonwhites. The rapid growth of Asian and Latino communities has stimulated hostile attitudes of natives toward these groups; the hostile reaction has been very pronounced among native blacks, who, even if themselves the victims of oppression, are also native-born Americans (Harwood, 1986).

As Asian and Latino immigrants move into traditionally black areas, the potential for interethnic urban conflict has increased. In Los Angeles, blacks have collided with Latinos, charging that these newcomers underbid them at work and outbid them for housing (Oliver & Johnson, 1984). Similarly, in

Miami, blacks displayed a nativist resentment of Cuban immigrants. In May 1980, black-Cuban competition for jobs and housing was a background precipitant of a two-day racial riot in Miami's Liberty District, a solidly black residential neighborhood (Sheppard, 1980). Asian immigrants have also collided with blacks. In particular, the growing presence of Korean businesses in black neighborhoods in Baltimore, Philadelphia, Washington, D.C., New York City, and Los Angeles has fueled black anger, at times leading to racial violence (Kim, 1981; Light & Bonacich, 1988, chapter 12; Cheng & Espiritu, 1989).

Recent studies of conflicts among nonwhite minorities suggest that intergroup antagonism revolves around economic issues, particularly job competition (Kim, 1981; Light, 1988). Structural changes in the economy in the past few decades partly set the stage for these urban conflicts. The automation of assembly line work, the growth of large multinationals, and the movement of industry to the suburbs and, more recently, outside of the country, had eliminated many entry-level, unskilled jobs traditionally held by nonwhites (Bluestone & Harrison, 1982; Oliver & Johnson, 1984; Wilson, 1978). The influx of new immigrants into the cities, particularly of undocumented aliens, has exacerbated the competition over scarce jobs. Because of their ethnic and cultural differences, the new immigrants are perfect targets for the displaced hostility of the traditional inhabitants of urban areas (Oliver & Johnson, 1984, p. 86). In the context of declining opportunities and increasing competition for economic subsistence in urban America, the potential for conflict between nonwhite groups is ever increasing.

Conflict among nonwhite minorities is also intensified by their unfavorable attitudes toward one another. In a secondary analysis of a 1980 *Los Angeles Times* survey,[1] Oliver and Johnson (1984, p. 84) reported that "there exists a strand of ethnic antipathy between blacks and Latinos that, when fueled by other factors, could lead to conflict between these groups." In the case of blacks and Koreans, a study of the social distance of black college students finds that these students place Koreans at the bottom of their social distance scale (Schaefer, 1987, p. 31). Koreans also harbor unfavorable attitudes toward blacks, partly derived from their negative encounter with American black soldiers in the Korean war, and partly from the negative depiction of blacks in American movies, television shows, and newspaper accounts (Lim, 1982).

Conflicts among nonwhite minorities illustrate the theoretical tension between prejudice and resource competition theories. Motives are often mixed so that economically motivated acts may also carry a racist message and vice versa. In a review of the etiology of intergroup conflicts, Light

(1983, chapter 13) posits that prejudice and resource competition theories are not necessarily incompatible. Instead of choosing between prejudice and resource competition, Light proposes that intergroup antagonism is strongest when unfavorable attitudes and resource competition coincide, and weakest when attitudes are favorable and competition is nil.

For the most part, research on interethnic urban relations has approached the topic through an analysis of conflict. We find few studies that focus on the cooperative efforts among nonwhite minorities. Prompted by the hope of enhanced clout, the country's two fastest growing minority groups, Asians and Latinos, occasionally have forged new political alliances. In 1987, Hispanic and Asian groups in New York teamed together to promote multilingualism. In San Francisco, Asians and Latinos formed immigration advocacy groups to educate people about their rights under the 1987 federal immigration bill (Fong, 1987). Given current population trends, the 10 largest U.S. cities will be more than 50% minority within a generation (Falcon, 1988, p. 172). This incipient demographic preponderance has encouraged Asians, blacks, and Latinos to forge political coalitions intended to translate population preponderance into political preponderance. In Los Angeles, noting the decrease in the white proportion of the county population, some blacks, Latinos, and Asians have called for political alignment, dubbing themselves the "new majority" (Ridley-Thomas, Pastor, & Kwoh, 1989). Due to their desire for solidarity with Latino colleagues, and the heavy Latino presence in many of their districts, black congressmen unanimously opposed employer sanctions—even though, fearing job displacement, many black advocacy groups supported the sanctions (Simcox, 1988, p. 6).

In sum, in light of the emerging numerical dominance of nonwhite groups in some urban areas, researchers need to be more attentive to contentious as well as cooperative relations among the minority groups.

SOCIOECONOMIC ADAPTATION

For many years, the assimilation model framed the study of ethnic socioeconomic progress. Despite the many strands that pervade this literature, its basic insight is that the market economy and democratic political institutions will lead to an eventual elimination of ascriptive ties—and thus a narrowing of socioeconomic disparities (Gordon, 1964; Park, 1950; Sowell, 1981). This assimilation hypothesis implies that, in the course of several generations, immigrant groups pass from initial economic hardship to eventual socioeconomic attainment arising from increasing knowledge of American "core" culture. In general, the assimilation model has been effective in explaining the adaptation of European ethnic communities. Although life was harsh for European immigrants upon arrival, their descendants have done relatively well in the United States. European ethnic origin has not been a significant barrier to socioeconomic achievement (Duncan & Duncan, 1968; Lieberson, 1980; Lieberson & Waters, 1988). In the last 20 years, however, the field of ethnic stratification has developed some new approaches. Some of these approaches originate from innovative studies of the old problem of immigrant absorption; others arise out of new empirical realities that require new thinking.

To explain ethnic stratification, the assimilation model has traditionally emphasized human capital variables, the set of learned skills that are rewarded by wages earned in the labor market (Chiswick, 1978, 1979; Featherman & Hauser, 1978). Education is often the primary indicator of human capital, along with work experience, on-the-job training, time of entry into the United States, and language skills. The premise behind this line of inquiry is straightforward: Immigrants may initially receive less than equal labor market outcomes because of the inferior skills or qualifications that they bring to the market (Hirschman, 1982, p. 471). When their human capital levels have reached those of native workers, however, immigrants and their descendants cease to be disadvantaged in the labor market.

Challenging the narrow assumptions of the human capital approach, recent research emphasizes the effects of the structural context on socioeconomic mobility (Lieberson, 1980; Wilson, 1978). While internal group characteristics are important determinants of economic success, they must be considered in light of the different economic, political, and developmental dynamics of particular urban areas in which immigrants enter the labor market (Morawska, 1990, p. 196). Some studies point out that differences in economic achievements are often the product of different economic opportunities. For example, when European immigrants were arriving in large numbers in the late 19th century, America was becoming an urban-industrial society. The growth of modern industry in northeastern and northcentral cities absorbed the millions of South and East European immigrants to the United States. In contrast, by the time blacks and Latinos appeared on the urban scene, opportunities for unskilled labor had declined drastically (Oliver & Johnson, 1984, p. 62). As the National Advisory Commission on Civil Rights (1973, p. 99) stated, "The Negro migrant, unlike the immigrant migrant, found little opportunity in the city; he had arrived too late, and the unskilled labor he had to offer was no longer needed."

Other structural studies emphasize institutional barriers. In a comparison of the economic progress of blacks with Southern, Central, and Eastern Europeans, Lieberson (1980) contrasted the reception of black migrants with that of the new Europeans in the northern cities. He reported that blacks faced greater barriers toward participation in every institutional area; most significant was the hostility of labor unions toward blacks. Since blacks were denied entry into related areas of the economy and society (neighborhoods, schools), they fell behind the progress made by European groups during both the pre- and post-World War II eras.

Research in recent years has also indicated the limitations of continuing to treat the absorbing social structure as a homogeneous given to which immigrants homogeneously respond. Immigrants do not compete in a single labor market; instead, they enter metropolitan economies divided into the monopoly, competitive, and informal sectors (Bibb & Form, 1977; Bonacich, 1972; Light, 1983, chapter 13). Sociologists today argue that the majority economy has bifurcated into at least two distinct economies. The primary sector is characterized by monopolies and oligopolies. In this favored labor market, wages are high, union security prevails, jobs are secure, and career hierarchies exist. In contrast, firms in the secondary sector are small, non-union, subject to fierce competition, and likely to fail. Employment in this sector means low wages, unattractive or even dangerous working conditions, and limited mobility opportunities (Briggs, 1975; O'Connor, 1973).

Valid as far as it went, dual labor market theory oversimplified the economic situation of immigrants because it overlooked a third sector of the economy, the immigrant business sector. As currently understood, the primary and secondary sectors of the general labor market coexist with an immigrant business sector in which immigrants work as employees of coethnics or as entrepreneurs (Bonacich & Modell, 1980; Light, 1972; Light & Bonacich, 1988; Portes & Bach, 1985). The size of the immigrant sector is variable as is its formality. That is, some immigrant groups have large business sectors, some small. In some groups, the immigrant business sector is mainly formal, in others mainly informal. Either way, immigrants find employment in the general labor market or in the immigrant business sector; they switch back and forth between the sectors, and they trade off the costs and benefits of employment in either sector. Much current literature deals with those trade-offs. Some writers have argued that immigrant employees earn higher returns as employees of coethnic entrepreneurs (Portes & Bach, 1985); others have argued the reverse (Sanders & Nee, 1987). Results thus far indicate that sometimes one, sometimes the other pattern prevails depending upon locality, gender, and specific group affiliations.

In sum, above and beyond the human capital possessed by individuals, their differential insertion into the labor market means access to unequal opportunities and results in unequal socioeconomic attainments (Morawska, 1990, p. 199). The economic incorporation of immigrants thus depends upon the proportion who enter the secondary, primary, and immigrant business sectors.

The heterogeneity of the new waves of immigrants and the diversity of their socioeconomic progress has also challenged the assumption that socioeconomic mobility is a uniform process (Hirschman, 1982, p. 476). Compared to earlier inflows of primarily rural and unskilled workers, recent newcomers comprise a broader mix of socioeconomic backgrounds, drawing on both ends of the educational and occupational spectrums (Massey, 1981, p. 63). Today's immigrants include first generation millionaires and topflight engineers and scientists; at the other extreme, they also include destitute refugees and undocumented workers. Because of their illegal status, the undocumented cannot even take the first step toward assimilation (Portes & Rumbaut, 1990, p. 8).

Unlike earlier rural migrants from Europe, mostly uprooted ex-peasants, today's Asian, Middle Eastern, Soviet, and Latin American immigrants frequently represent a select and high-status group from their home country's population: they very commonly have college degrees, money in the bank, and sophisticated occupational and business skills (Light, 1983, p. 323). During the last decade, when professionals and technicians amounted to no more than 18% of the American labor force, professionals represented about 25% of the total number of active immigrants (Portes & Rumbaut, 1990, p. 10). In particular, recent immigrants from Asian countries are predominantly highly educated, and well-trained professional or skilled persons from urban areas. For example, professionals represent almost half of occupationally active immigrants from India and almost a third of those from Taiwan (Light & Bonacich, 1988; Minocha, 1987; Portes & Rumbaut, 1990). As Gardner, Robey, and Smith (1985, p. 32) stated, "For most recent Asian immigrants, the old U.S. image of immigrants filling only the lowest rungs on the occupational ladder no longer applies."

Awarding preference to highly trained workers in short supply, the 1965 Immigration and Naturalization Act has encouraged the influx of professionals. But the immigration of professionals has also been linked to the structural imbalances in the world system. Like labor migration, professional migration is the product of externally induced imbalances in the sending countries. As an example, when the technical apparatus of advanced nations is reproduced in underdeveloped ones, these implanted institutions function more in accordance with the needs and requirements of the advanced nations than those of the receiving country. Hence the labor market demand for foreign-trained workers is often weak. These structural imbalances lead to migration to more advanced economies (Portes & Walton, 1981, pp. 36-39). The large immigration of Filipino medical professionals to the United States exemplifies this process. Furthermore, increasing interdependencies between developing countries and the United States create consumption standards based on those in the United States. These external standards become incorporated into normative expectations; however, these expectations seldom match the actual resources and possibilities of peripheral countries. These contradictions are most poignant for the better educated and more modernized sectors. Consequently, they are most susceptible to emigration (Portes & Bach, 1985, pp. 6-7).

Much research on ethnicity in the United States has focused on the adaptation and assumed assimilation of unskilled laborers to the dominant white social structure and culture (Cheng & Bonacich, 1984; Sassen, 1988). Drawing on the dual labor market perspective, this literature views recent immigrants as the latest entrants into the low-wage secondary occupations (Sassen-Koob, 1980). The economic profile of today's influx, however, suggests that immigrants are not uniformly directed to the secondary labor market. Instead, "a significant number and proportion of those arriving today are directed to the primary sector of the economy" (Morawska, 1990, p. 201). Although case studies are regrettably few, Monterey Park, California, illustrates the impact that wealthy immigrants can have on American communities. A near-suburb of Los Angeles, Monterey Park is the commercial and financial hub of Southern California's Asian population. Dubbed "America's first suburban Chinatown," Monterey Park has undergone rapid and complete ethnic transformation in the last 20 years (Arax, 1987). As recently as 1960, the city was 85% white. In 1988, after an influx of wealthy professionals and businessmen from Taiwan and Hong Kong, the Chinese had become the largest ethnic group in Monterey Park, totaling more than 50% of the suburban city's 61,000 residents. Instead of adapting to and adopting the dominant Anglo-American culture, the Chinese newcomers are building an American city in their own tradition. And they have the capital to do so. In 1985, *Forbes* magazine estimated that $1.5 billion had been deposited in Monterey Park financial institutions during the year, or about $25,000 for every man, woman, and child. The Chinese own an estimated two thirds of the property and business in the city. As a result, Monterey Park has acquired a new ethnic face: English business signs have been overshadowed by blocks of uninterrupted Chinese-language signs (Takaki, 1989, pp. 425-426).

Political exiles represent another important component of the new immigration. Although labor immigrations have continued to be the norm in the United States during the last two decades, and Mexicans are the largest component of this migration, labor migrations have been accompanied by increasingly numerous inflows of refugees, primarily from Iran, Lebanon, the Soviet Union, Cuba, and Southeast Asia. Although more research is needed on refugees, preliminary studies indicate that the mobility patterns of refugees differ from those of other immigrants. In a study of the economic progress of 11 different immigrant groups, Chiswick (1979) concluded that refugees face the steepest barriers in achieving economic success. Because their migration is influenced by political factors rather than labor market success, refugees are less likely to be favorably self-selected than economic migrants. They are also less likely to have acquired readily transferable skills and are more likely to have made investments (in training and education) specific to their country of origin (Chiswick, 1979; Montero, 1980). The study of refugees is further complicated by the socioeconomic bipolarity of the groups. For both Cuban and Southeast Asian refugees, the early cohorts were disproportionally from well-educated and professional backgrounds while the later arrivals were from less-educated and working-class origins (Portes, Clark, & Bach, 1977; Stein, 1979). This broad mix of socioeconomic backgrounds indicates that refugees can be incorporated in any of the three economic sectors: the primary labor market, the secondary sector, or the ethnic economy.

SOCIOCULTURAL ADAPTATION

The assimilationist perspective informs most of the classic studies of immigrants in the United States (Child, 1943; Handlin, 1951; Wittke, 1952). Treating ethnicity as an import from abroad, not a product of domestic social conditions, assimilationists thought that ethnic heritages would not long survive in this country's inhospitable environment. Assimilationists suppose that intergroup contact promotes acculturation and, after sufficient time, even assimilation, a merging of cultural values, primary group memberships, and identities (Yinger, 1985, p. 154). This process of boundary reduction has been most evident among white ethnic groups. Lieberson and Waters (1988) reported that in 1980, the once large socioeconomic and cultural gaps between white ethnic groups had all but disappeared; at the same time, intermarriages among white ethnics had increased. In contrast, "for whatever the cause, a European-non-European distinction remains a central division in the society" (Lieberson & Waters, 1988, p. 248).

Recent studies of immigration and ethnicity have moved away from the once-dominant assimilation model. It is not that assimilation has not occurred. Much evidence of assimilation exists. Even decades of intergroup contacts, however, have not eventuated in the abandonment of distinct ethnic identities and social networks. Traditional sources of solidarity proved far more resilient than had been expected. Ethnic distinctions still thrive in urban areas, particularly as a basis for political organization (Bonacich & Modell, 1980; Glazer & Moynihan, 1963; Novak, 1972). As a consequence, the focus of research in recent years has shifted toward explaining the formation and persistence of ethnicity in contemporary societies with attention paid to intragroup differences in rate of assimilation.

While the assimilation model predicts that increased contact among members of different ethnic groups weakens group boundaries, some recent research suggests that increased intergroup contacts actually promote and strengthen ethnic solidarity, the popular awareness of ethnic connections and heritage. Contacts increase awareness of intergroup differences in income, wealth, power, living standards, and prestige. The invidious awareness of inequality sharpens ethnic bonds and increases underdog mobilization around the symbols of a common ethnicity (Portes, 1984). Naturally, cultural heritage is a major source of group identity; however, another important source is the collective perception of injustice based on a fundamental and persistent condition of group inequality (Barrera, 1979). As Hechter (1975) reported, when individuals are assigned to subordinate economic and social roles on the basis of observable cultural traits or markers, their political demands will largely be formulated in ethnic terms. Even when those in subordinate positions do not initially regard themselves as being alike, "a sense of identity gradually emerges from a recognition of their common fate" (Shibutani & Kwan, 1965, p. 208). Therefore, ethnicity is not simply a primordial attachment to a cultural heritage, already present upon arrival. Rather, ethnicity emerges from and varies with the absorption process in American cities. It is, in that sense, a product of American social conditions, not an import. The imputation of a common ethnicity by the core society and its use to justify their exploitation bring together immigrants who may have shared only the most tenuous ties in the old country (Yancey, Ericksen, & Juliani, 1976).

Ethnicity also survives because it is a valuable tool for the protection or enhancement of status. The functional advantages of ethnicity range from "the moral and material support provided by ethnic networks to political gains made through ethnic bloc voting" (Portes & Bach, 1985, p. 24). In other words, ethnic groups are not only sentimental associations of persons sharing affective, primordial ties; they are also interest groups. Bonacich and Modell's study of Japanese Americans (1980) provides a clear example of the interest basis of ethnic solidarity. To explain ethnic survival, the authors investigated the relationship between ethnicity and economic interests of Japanese Americans. Using the middleman minority model (Bonacich, 1973), they showed that those who worked in the ethnic economy were consistently more "Japanese" than those who held jobs in the general labor market. While the owners of small businesses relied on ethnic affiliation for human and financial resources, those who moved into the corporate economy had less reason to retain ethnic ties to fellow Japanese. Thus the authors predicted that without common and distinctive class interests, it would be increasingly difficult for ethnic groups to keep members from disappearing into the mainstream. Bonacich and Modell's study calls attention to the fact that ethnicity is a variable in need of explanation. In so doing, it raises important research questions: Under what conditions will ethnic sentiments be invoked and under what conditions will they subside as a major axis of social organization?

Recent research suggests that differences in the patterns of adaptation among immigrants correspond to differences in the form of their labor market incorporation. For those who work in an ethnic economy, acculturation is slowest. Once an enclave economy has fully developed, it is possible for a newcomer to find work, education, recreation, access to health care, and other essential services without leaving the enclave (Portes & Manning, 1986). This institutional completeness allows ethnic entrepreneurs to move ahead economically, despite very limited knowledge of the language and culture of the host society (Light, 1972: Portes & Bach, 1985). The strong ethnic networks within these economic enclaves reinforce the culture of origin and strengthen ethnic ties.

On the other hand, the acculturation of professionals such as physicians, engineers, and scientists approximates the predictions of the assimilation model. Primary sector immigrants are among the most assimilated. Professional and technical immigrants tend to rely less on the assistance of preexisting ethnic communities than do working-class immigrants. The skilled immigrants depend on their own skills and qualifications. Consequently, they are more dispersed among and within metropolitan regions than are working-class immigrants in other modes of incorporation (Portes & Rumbaut, 1990, p. 33). For example, dispersion of population and invisibility of business is the hallmark of the wealthy and well-educated Iranian population in Los Angeles (Sabagh & Light, 1985). Similarly, large numbers of upper and middle-class Asians often have no tie with residents of the older coethnic communities that await them upon arrival. Chinese from Taiwan do not even speak the same dialect as do Chinese from Canton, the earlier settlers. Instead, the new immigrants "have scattered into white communities to pursue occupational interests and initiate business ventures which have traditionally been dominated by whites" (Kuo, 1979, pp. 286-287).

The experience of professional and moneyed immigrants provides a theoretically interesting deviant case of immigrants without a visible ethnic community. This case is deviant because it contradicts the well-supported generalization in more than half a century of American social research that immigrants cluster in residential and commercial enclaves in order to cope with a strange and often hostile urban environment. Thus while peasants and impoverished immigrants require collectivism to cope, educated and wealthy immigrants need less community support because they have skills and resources that permit them to operate independently in the general labor market (Sabagh & Light, 1985, p. 4). The occupational bimodality of today's immigrants provides researchers with a good opportunity to conduct comparative studies of the adaptation process of different types of immigrants.

Although professional immigrants are among the most rapidly assimilated, they are also among the most likely to retain relations with their country of origin. For example, well-to-do Asian Indians continue to maintain strong ties and to find spouses for themselves and their children in India (Lopez, 1988). Their financial resources and ease of international transportation make it possible for professional immigrants to make periodic visits to their home country to keep up with personal and social relations. Because they have the means to do so, these immigrants often own homes in both the United States and in their country of origin, with not only themselves but also other members of the family moving back and forth several times a year (Cheng, 1986; Portes & Rumbaut, 1990). These return trips and easier communication with family and friends at home serve to keep alive the identifications and loyalties into which they were socialized (Portes & Rumbaut, 1990, pp. 109-110). This continuing link with the sending country can affect the group's development of an ethnic identity in the receiving country. As Lopez (1988) reported, "the general and specific conflicts in South Asia militate against the development of strong pan-Indian identity and association in the United States."

Traditionally, scholars view migration as a simple unidirectional movement: as immigrants assimilate the culture of the new environment, they diminish their involvement in the home country. The continuing link between today's immigrants and their countries of origin calls attention to the fluidity of the process of migration. Although back and forth flows between homeland and destination are not new, they are accentuated both by the flow of the highly trained and by the increasing economic connections between nations (Cheng, 1986). In recent years, developing countries have become increasingly interested in the role that their overseas compatriots can play in investment and in the transfer of technology. Governments want their nationals abroad to invest in the economic development of the mother country. For example, the Chinese and Indian governments now actively work with their migrant business owners and professionals to promote investment and technology transfers (Weiner, 1985, p. 452). By becoming agents of linkage of economies between their home and receiving countries, today's immigrants become linchpins of global restructuring.

Along the same line, today's migration patterns include many multilateral moves. Many post-1965 immigrants do not simply move from a sending country to a receiving country but they also undergo multiple displacements. For example, large numbers of ethnic Chinese immigrants are secondary and even tertiary movers. Initially refugees from the People's Republic of China, Chinese immigrants usually emigrated from a second point of departure such as Hong Kong or Taiwan rather than directly from mainland China. Others began their journey in China, lived in Hong Kong for a considerable portion of their lives, and then moved to Canada before coming to the United States (Cheng, 1986; Nee & Nee, 1976; Takaki, 1989). Given this empirical reality, researchers need to differentiate between the ethnicization experiences of the direct and secondary or tertiary migrants (Bhachu, 1985; Espiritu, 1989; Viviani, 1984). In a study of Chinese Vietnamese who migrated from China to Vietnam and then to the United States, Espiritu (1989) suggested that their common experiences in Vietnam have given Chinese Vietnamese skills that they utilize in this country. Since these "twice minorities" were already part of an established community before resettlement in the United States, they have been able to reproduce community ties and to establish ethnic institutions rapidly upon arrival. Their skills have helped them to establish themselves much more rapidly than the direct migrants who lacked the same expertise, linguistic facility, and communications network to develop community structures at the same pace.

The arrival of large numbers of new immigrants has also produced an ethnic flux among the established ethnic groups in American cities. The history of Asian immigration to the United States provides a good example. For some Asian groups, such as the Japanese, immigration has all but ceased since the first wave in the 19th and early 20th centuries. Without an influx of new immigrants from Japan, Japanese Americans have become a predominantly native-born population. In 1980, 72% were citizens by birth. On the other hand, other Asian groups have been inundated by new immigrants. In 1980, 63% of the Chinese, 66% of the Filipinos, and close to 82% of the Koreans were foreign born. For these groups, a period of more than 50 years separates the first significant wave of immigration from the contemporary, second wave. Due to the wide gap in their time of arrival and the differences in their socioeconomic characteristics, the recent arrivals often clash with their counterparts who came decades ago (Cheng, 1986; Kuo, 1979).

Continuing immigration has also made it difficult to maintain Asian panethnicity. Panethnicity is the coalition of national-origin defined ethnic groups into alliances with kindred groups derived, however, from another nation. For the most part, panethnic organization emerges because ethnic groups recognize that their interests have not been well served by subgroup divisions nor by piecemeal assimilation into American society (Light, 1981, p. 79). Variation on immigration patterns across Asian groups has affected panethnic development. Asian groups that have been in the United States for similar lengths of time are more likely to share similar concerns and interests. In contrast, groups differing markedly in generational composition lack a basis for cooperation (Lopez & Espiritu, 1990). Moreover, native Asians may sense competition from Asian immigrants and the newcomers may feel dominated by the native born. In Los Angeles, the competition for jobs, money, and power among Asian Americans has occasionally been bitter, with some groups, particularly Filipinos and other big groups, choosing to pursue their interests outside the pan-Asian framework (Espiritu, 1990, chapter 3.)

CONCLUSION

In the past, immigrants entered American society at the bottom and worked their way up the socioeconomic hierarchy in the course of generations. At least in the case of whites, this movement was accompanied by acculturation and assimilation such that barriers between groups came down. Acculturation and assimilation are still reducing barriers. This is especially true of foreign professional and technical workers who make up a big share of current immigration to the United States.

But recent research has been impressed with the extent to which American society generates ethnicity rather than simply dissolving it. Partially as a result of urban restructuring, a process that drastically realigns employment and residence patterns, America's immigrants confront a more complex political, economic, and social landscape than did earlier generations of immigrants. These complex and often threatening conditions encourage them to band together around common linguistic and cultural identities in order to influence their life chances and living conditions. Additionally, the proliferation of groups, and the appearance of historic newcomers like Vietnamese, Iranians, and Armenians, have promoted divisive conflicts that pit one immigrant and ethnic minority against another in competition for scarce jobs, housing, political influence, and even parking spaces.

New conditions beget new theoretical approaches. These build upon those of earlier generations, but modify them in the light of current experience. In treating the dilemmas and problems of contemporary immigrants, we must draw upon the assimilation literature that organized and explained the experiences of their predecessors. At the same time, sociologists have had to modify that perspective to take account of the new pluralism in a new and restructured urban environment.

NOTE

1. This survey ($N = 1,295$) oversampled blacks and Mexicans in Watts and East Los Angeles, the residential centers of both populations.

REFERENCES

Allport, G. R. (1958). *The nature of prejudice*. Boston: Beacon.

Arax, M. (1987, April 6). Monterey Park: Nation's first suburban Chinatown. *Los Angeles Times*.

Barrera, M. (1979). *Race and class in the southwest: A theory of racial inequality*. London: University of Notre Dame Press.

Bhachu, P. (1985). *Twice migrants: East African Sikh settlers in Britain*. London: Tavistock.

Bibb, R., & Form, W. H. (1977). The effects of industrial, occupational and sex stratification in blue-collar markets. *Social Forces, 55*, 974-996.

Blauner, R. (1969). Internal colonialism and ghetto revolt. *Social Problems, 16*, 393-408.

Bluestone, B., & Harrison, B. (1982). *The deindustrialization of America: Plant closings, community abandonment, and the dismantling of basic industry*. New York: Basic Books.

Bonacich, E. (1972). A theory of ethnic antagonism: The split labor market. *American Sociological Review, 37*, 547-559.

Bonacich, E. (1973). A theory of middleman minorities. *American Sociological Review, 38*, 583-594.

Bonacich, E. (1976). Advanced capitalism and black-white relations in the United States: A split labor market interpretation. *American Sociological Review, 41*, 34-51.

Bonacich, E., & Modell, J. (1980). *The economic basis of ethnic solidarity: A study of Japanese Americans*. Berkeley: University of California Press.

Briggs, V. M. (1975). Illegal aliens: The need for a more restrictive border policy. *Social Science Quarterly, 3*, 477-484.

Briggs, V. M. (1985). Employment trends and contemporary immigration policy. In N. Glazer (Ed.), *Clamor at the gates: The new American immigration* (pp. 135-160). San Francisco: Institute for Contemporary Studies.

Cheng, L. (1986). *The sociology and historiography of immigration: An Asian American perspective.* Unpublished manuscript.

Cheng, L., & Bonacich, E. (Eds.). (1984). *Labor immigration under capitalism: Asian workers in the United States before World War II.* Berkeley: University of California Press.

Cheng, L., & Espiritu, Y. L. (1989). Korean businesses in black and Hispanic neighborhoods: A study of intergroup relations. *Sociological Perspectives, 32,* 521-534.

Child, I. (1943). *Italian or American? The second generation in conflict.* New Haven, CT: Yale University Press.

Chiswick, B. (1978). The effect of Americanization on the earnings of foreign born men. *Journal of Political Economy, 86,* 897-921.

Chiswick, B. (1979). The economic progress of immigrants: Some apparently universal patterns. In W. Fellner (Ed.), *Contemporary economic problems* (pp. 357-399). Washington, DC: American Enterprise Institute.

Cummings, S. (1980). White ethnics, racial prejudice, and labor market segmentation. *American Journal of Sociology, 85,* 938-950.

Dinnerstein, L., Nichols, R. L., & Reimers, D. M. (1990). *Natives and strangers: Blacks, Indians, and immigrants in America.* New York: Oxford University Press. (Original work published 1979)

Duncan, B., & Duncan, O. D. (1968). Minorities and the process of stratification. *American Sociological Review, 33,* 356-364.

Espiritu, Y. L. (1989). Beyond the "boat people": Ethnicization in American life. *Amerasia Journal, 15*(2), 49-67.

Espiritu, Y. L. (1990). *Cooperation and conflict: Panethnicity among Asian Americans.* Doctoral dissertation, University of California, Los Angeles.

Falcon, A. (1988). Black and Latino politics in New York City: Race and ethnicity in a changing urban context. In F. C. Garcia (Ed.), *Latinos and the political system* (pp. 171-194). Notre Dame, IN: Notre Dame University Press.

Featherman, D. L., & Hauser, R. M. (1978). *Opportunity and change.* New York: Academic Press.

Fong, T. (1987, July 9). Asian small business growth becomes lightning rod for anti-Asian sentiment. *East West.*

Gardner, R. W., Robey, B., & Smith, P. (1985). *Asian Americans: Growth, change, and diversity* (Population Bulletin No. 40:4). Washington, DC: Population Reference Bureau.

Glazer, N., & Moynihan, P. (1963). *Beyond the melting pot: The negroes, Puerto Ricans, Jews, Italians, and Irish of New York City.* Cambridge, MA: MIT Press.

Gordon, M. (1964). *Assimilation in American life: The role of race, religion, and national origins.* New York: Oxford University Press.

Handlin, O. (1951). *The uprooted.* Boston: Little, Brown.

Harwood, E. (1986). American public opinion and U.S. immigration policy. *Annals of the American Academy of Political and Social Science, 487,* 201-212.

Hechter, M. (1975). *Internal colonialism: The Celtic fringe in British national development, 1536-1966.* Berkeley: University of California Press.

Hirschman, C. (1982). Immigrants and minorities: Old questions for new directions in research. *International Migration Review, 16,* 475-490.

Kim, I. (*1981*). *New urban immigrants: The Korean community in New York.* Princeton, NJ: Princeton University Press.

Kinder, D. R., & Sears, D. O. (1981). Prejudice and politics: Symbolic racism versus racial threats to the good life. *Journal of Personality and Social Psychology, 40,* 414-431.

Kuo, W. H. (1979). On the study of Asian Americans: Its current state and agenda. *Sociological Quarterly, 20,* 279-290.

Lieberson, S. (1980). *A piece of the pie: Black and white immigrants since 1880.* Berkeley: University of California Press.

Lieberson, S., & Waters, M. C. (1988). *From many strands: Ethnic and racial groups in contemporary America.* New York: Russell Sage.

Light, I. (1972). *Ethnic enterprise in America.* Berkeley: University of California Press.

Light, I. (1981). Ethnic succession. In C. Keyes (Ed.), *Ethnic change* (pp. 54-86). Seattle: University of Washington Press.

Light, I. (1983). *Cities in world perspective.* New York: Macmillan

Light, I. (1988). Los Angeles. In M. Dogan & J. Kasarda (Eds.), *The metropolis era: Vol. 2. Mega-cities* (pp. 56-96). Newbury Park, CA: Sage.

Light, I., & Bonacich, E. (1988). *Immigrant entrepreneurs: Koreans in Los Angeles 1965-1982.* Berkeley: University of California Press.

Lim, H. (1982). Acceptance of American culture in Korea: Patterns of cultural contact and Koreans' perception of American culture. *Journal of Asiatic Studies, 25,* 25-36.

Lopez, D. (1988). *The organization of ethnicity: Asian Indian associations in the United States.* Research proposal submitted to the Social Science Research Council, Los Angeles.

Lopez, D., & Espiritu, Y. L. (1990). Panethnicity in the United States: A theoretical framework. *Ethnic and Racial Studies, 13* (20), 98-224.

Massey, D. S. (1981). Dimensions of the new immigration to the United States and the prospects for assimilation. *Annual Review of Sociology, 7,* 57-85.

Middleton, R. (1976). Regional differences in prejudice. *American Sociological Review, 41,* 94-117.

Minocha, U. (1987). South Asian immigrants: Trends and impacts on the sending and receiving societies. In J. T. Fawcett & B. Carino (Eds.), *Pacific bridges* (pp. 347-374). New York: Center for Migration Studies.

Montero, D. (1980). *Vietnamese Americans: Patterns of resettlement and socioeconomic adaptation in the United States.* Boulder, CO: Westview.

Morawska, E. (1990). The sociology and historiography of immigration. In V. Yans-McLaughlin (Ed.), *Immigration reconsidered: History, sociology, and politics* (pp. 187-238). New York: Oxford University Press.

Muller, T., & Espenshade, T. J. (1985). *The fourth wave: California's newest immigrants.* Washington, DC: Urban Institute Press.

National Advisory Commission on Civil Rights. (1973). Comparing the immigrant and negro experiences. In E. G. Epps (Ed.), *Race relations* (pp. 99-101). New York: Winthrop.

Nee, V. G., & Nee, B. de B. (1976). The emergence of a new working class. In E. Gee (Ed.), *Counterpoint: Perspectives on Asian America* (pp. 376-379). Los Angeles: UCLA Asian American Studies Center.

Novak, M. (1972). *The rise of the unmeltable ethnics.* New York: Macmillan.

O'Connor, J. (1973). *The fiscal crisis of the state.* New York: St. Martin's.

Oliver, M. L., & Johnson, J. H., Jr. (1984). Inter-ethnic conflict in an urban ghetto: The case of blacks and Latinos in Los Angeles. *Research in Social Movements, Conflict and Change, 6,* 57-94.

Park, R. (1950). *Race and culture*. Glencoe, IL: Free Press.

Piore, M. J., & Sabel, C. F. (1984). *The second industrial divide: Possibilities for prosperity*. New York: Basic Books.

Portes, A. (1984). The rise of ethnicity: Determinants of ethnic perceptions among Cuban exiles in Miami. *American Sociological Review, 49,* 383-397.

Portes, A., & Bach, R. L. (1985). *Latin journey: Cuban and Mexican immigrants in the United States*. Berkeley: University of California Press.

Portes, A., Clark, J. M., & Bach, R. L. (1977). The new wave: A statistical profile of recent Cuban exiles in the United States. *Cuban Studies, 7,* 1-32.

Portes, A., & Manning, R. (1986). The immigrant enclave: Theory and empirical examples. In J. Nagel & S. Olzak (Eds.), *Competitive ethnic relations* (pp. 46-64). Orlando, FL: Academic Press.

Portes, A., & Rumbaut, R. G. (1990). *Immigrant America: A portrait*. Berkeley: University of California Press.

Portes, A., & Walton, J. (1981). *Labor, class, and the international system*. San Diego, CA: Academic Press.

Ridley-Thomas, M., Pastor, M., & Kwoh, S. (1989, October 12). The "new majority" wants its share. *Los Angeles Times*, p. 87.

Sabagh, G., & Light, I. (1985). *Emergent ethnicity: Iranian immigrant communities*. A grant application submitted to the National Science Foundation.

Sanders, J. W., & Nee, V. (1987). Limits of ethnic solidarity. *American Sociological Review, 52,* 745-767.

Sassen, S. (1988). *The mobility of labor and capital: A study in international investment and labor flow*. Cambridge, UK: Cambridge University Press.

Sassen-Koob, S. (1980). Immigrant and minority workers in the organization of the labor process. *Journal of Ethnic Studies, 1,* 1-34.

Schaefer, R. T. (1987). Social distance of black college students at a predominantly white university. *Social Science Research, 72,* 30-31.

Sheppard, N. (1980, May 23). Miami's black have "nothing to lose." *The New York Times*, II, p. 4.

Shibutani, T., & Kwan, K. M. (1965). *Ethnic stratification*. New York: Macmillan.

Simcox, D. (1988). Overview—A time of reform and reappraisal. In D. Simcox (Ed.), *U.S. immigration in the 1980s: Reappraisal and reform* (pp. 1-63). Boulder, CO: Westview.

Smith, M. P., & Feagin, J. R. (Eds.). (1987). *The capitalist city: Global restructuring and community politics*. New York: Basil Blackwell.

Soja, E., Morales, R., & Wolff, G. (1983). Urban restructuring: An analysis of social and spatial change in Los Angeles. *Economic Geography, 59,* 195-230.

Sowell, T. (1981). *Ethnic America*. New York: Basic Books.

Stein, B. (1979). Occupational adjustment of refugees: The Vietnamese in the United States. *International Migration Review, 13*(1), 25-45.

Takaki, R. (1989). *Strangers from a different shore: A history of Asian Americans*. Boston: Little, Brown.

Viviani, N. (1984). *The long journey: Vietnamese migration and settlement in Australia*. Carlton, Victoria: Melbourne University Press.

Weiner, M. (1985). On international migration and international relations. *Population and Developmental Review, 11,* 441-455.

Wilson, W. J. (1978). *The declining significance of race: Blacks and changing American institutions.* Chicago: University of Chicago Press.

Wittke, C. (1952). *Refugees of revolution: The German forty-eighters in America.* Philadelphia: University of Pennsylvania Press.

Yancey, W. C., Ericksen, E. P., & Juliani, R. N. (1976). Emergent ethnicity: A review and reformulations. *American Sociological Review, 41,* 391-403.

Yinger, M. J. (1985). Ethnicity. *Annual Review of Sociology, 11,* 151-180.

4

Racism and the City

ROBERT D. BULLARD
JOE R. FEAGIN

THE RETREAT
FROM RACIAL DISCRIMINATION

WHITE RACISM: THE 1960s

Briefly during the 1960s and early 1970s the civil rights movement, among other influences, pressured influential white scholars and observers to retheorize the racial problem in U.S. cities away from a moral and racial inferiority, blaming-the-black-victim perspective to one accenting structural discrimination crafted by whites. This scholarly and public discourse for a time used such theoretical concepts as *white racism, institutional racism, desegregation,* and *civil disobedience* to discuss the situation of black and white Americans. The 1968 Kerner Commission, appointed by President Lyndon B. Johnson to investigate the 1960's urban riots, theorized that "Our Nation is moving toward two societies, one black, one white—separate and unequal" (National Advisory Commission, 1968, p. 1). This presidential commission, 9 of whose 11 members were white, was headed by the white governor of Illinois. In the first pages of the widely cited report, the commissioners' conclusions were framed thusly:

> Discrimination and segregation have long permeated much of American life. . . .White society is deeply implicated in the ghetto. White institutions created it, white institutions maintain it, and white society condones it. . . . Race prejudice has shaped our history decisively; it now threatens to affect our future. White racism is essentially responsible for the explosive mixture which has been accumulating in our cities since the end of World War II. (National Advisory Commission, 1968, pp. 1, 5)

The conceptual language made extensive use of the ideas of white discrimination and racism and the body of the report documented discrimination in many urban institutions. In this way the distinguished white panel echoed the interpretations of prominent black analysts (see Carmichael & Hamilton, 1967).

THE RETREAT IN THE 1970s AND 1980s

This focus of theoretical and policy discourse on institutionalized racial discrimination did not last. Since the mid-1970s there has been a broad shift in how most white (and a few black) scholars, media commentators, politicians, and jurists describe and analyze the white-black problems in U.S. cities. Recent media articles have concluded the verdict of the Kerner Commission was much too harsh when it warned that the United States was moving "toward two societies." The authors of a *Newsweek* article argued that "mercifully, America today is not the bitterly sundered dual society the riot commission grimly foresaw" ("Black and White," 1988, p. 19). With the demise of a visible civil rights movement has developed a new language used not only by media commentators and politicians but also by dominant intellectuals to analyze American racial relations, one that accents such terms as *reverse discrimination, the black underclass, black family pathology,* and *the successful black middle class.* Scholars such as Nathan Glazer, Irving Kristol, Daniel Patrick Moynihan, Michael Novak, George Gilder, Thomas Sowell, Glenn Loury, Shelby Steele, and Charles Murray have developed some or all of these terms of analysis for looking at American racial relations. At the heart of their analyses is the refurbishment of the blaming-the-victim approach to interpreting racial problems.[1]

This new theoretization of urban racial problems has a twin focus. One central point is that racial discrimination, especially institutionalized discrimination, is in sharp decline; the other is that the main problems remaining for black Americans are matters of economics and personal morality, not race. Liberal scholars such as University of Chicago professor William J. Wilson, a black sociologist, have adopted this twin focus, accenting terms like urban *underclass.* Most of these scholars and similar journalistic analysts argue that economics and "class" (by which they mean social status) are now far more important as a problem for black Americans than are racial discrimination. In an influential book, *The Declining Significance of Race* (1978, p. 110-111.), Wilson contrasted the poverty-stricken conditions of a growing black underclass in central cities with the affluence and success of the black middle class moving out to the white suburbs. He argues that the growth of that middle class resulted from improving economic conditions and government policies such as equal employment programs. Wilson and other scholars view this affluent middle class not only as very successful but also as very substantially breaking down the major racial barriers in America in the areas of employment, public accommodations, and housing.

In the last decade the mass and intellectual media have developed the themes of black middle-class success and the declining significance of race in a steady drumbeat of articles on urban race relations. The major difference today compared to the 1960s, avers a *Newsweek* article, is the "emergence of an authentic black middle class, better educated, better paid, [and] better housed than any group of blacks that has gone before it" ("Black and White," 1988, p. 19). A particularly strong version of this argument about the demise of racial discrimination can be seen in a recent article by the literary editor of *The New Republic,* Leon Wieseltier. In this article Wieseltier (1989, pp. 19-20) writes about black leaders' views of the drug problem, views that accent white racism as a root cause. Wieseltier argues that this black perspective accenting white racism is "madness," that "in the memory of [racial] oppression, oppression outlives itself. The scar does the work of the wound. That is the real tragedy: that injustice retains the power to distort long after it has ceased to be real." The clear suggestion is that racial injustice and discrimination for the most part are no longer real.

This view of the declining significance of racial discrimination is shared by a majority of the white population. In reply to a 1980s Gallup poll question, "Looking back over the last ten years, do you think the quality of life of blacks in the U.S. has gotten better, stayed the same, or gotten worse?" More than half of the nonwhites (mostly blacks) said "gotten worse" or "stayed the same" ("Whites, Blacks," 1980, p. 10). Yet only a fifth of the white respondents answered in a similar way. Three quarters of the whites said "gotten better." A major 1988 Louis Harris survey (NAACP Legal Defense and Educational Fund, 1989, pp. 6-10) found sharp differences in the views of blacks and whites on job discrimination. In the case of executive-level jobs, 70% of the whites believed that blacks got fair treatment, while nearly two thirds of the black respondents felt they did not get a fair shake. In the case of white-collar jobs in general, 69% of the whites did not think blacks were discriminated against, but two thirds of the black respondents felt that blacks got paid less and were not treated equally. In the area of education this gap in evaluations was repeated.

From the perspective of most white media commentators, editors, jurists, politicians, and scholars, the problems of institutionalized racial discrimination and injustice have largely faded into the societal background; they are of no major relevance to problems facing black American today. The general white public has accepted the opinion makers' views that black Americans now get a fair shake.

RESURRECTING THE CONCEPT OF INSTITUTIONALIZED RACISM

ACCENTING RACE OVER "CLASS"

How accurate are these scholarly and public views? Do they capture the reality of life for black Americans today? Is antiblack discrimination no longer important in our society, beyond a few demented Ku Klux Klan members? Have middle-class blacks destroyed the racial barriers and achieved parity with middle-class whites? Are middle-class black Americans now equal in life chances with whites?

The realities of life for black Americans from the 1860s to the 1990s have been dramatically different from this whitewashed portrait of the declining significance of race and of most black Americans as doing well. While there have been some important remedial changes, institutionalized racism is still the condition faced daily by black Americans in all income levels. Both the poor and the middle class face major problems of racial discrimination. The badges of slavery have never been substantially eradicated. The Civil Rights Act of 1964, 1965, and 1968 made many exclusionary acts of discrimination illegal, but they did not end the broad array of blatant, subtle, and covert discrimination in jobs, housing, and education from the 1960s to the 1980s. The spectacle of institutionalized racism unwilling to die can be seen today in many examples, from which we take only a small sample:

(1) continued restrictions on black voting in urban and rural areas of the South;
(2) in the North and South most black children still attend de facto segregated schools;
(3) in the North and South most blacks are tried by all-white juries from which blacks have often been excluded in the selection process;
(4) in the North and South most blacks face discrimination in the job market, including promotion barriers;
(5) in the North and South most black families live in segregated residential areas with poor services or environmental threats;
(6) in the North and South most blacks seeking housing face informal discrimination by banks, real estate people, landlords, and homeowners.

While we cannot in this chapter address all these features of the racial scene that characterize urban (and rural) areas, we will address below the last two urban issues in some detail. Before providing empirical evidence that the U.S. city is grounded in institutionalized racism, we will first develop the concept of institutionalized racism.

INSTITUTIONALIZED RACISM

During the civil rights movements of the 1960s, several black and white scholars adopted the concept of *internal colonialism* to describe their model of racial adaptation and stratification in the United States. An emphasis on power and resource inequalities, particularly white-black inequalities, is at the heart of this model. Race is an independent factor linked to class but not reducible to class. Traditional assimilation-progress models are viewed as totally inadequate to the task of describing what has happened to black Americans, who were the only immigrant group brought into the emerging economic and political system initially by *force* as slaves. Subsequent black adaptation to the white system was along nonegalitarian lines, with continuing elements of coercion and systematic oppression. Internal black colonialism emerged out of European imperialism, and it is a system grounded in the sharp differentiation of white and black labor, in economic exploitation. Yet it was the color of black Americans that made it so easy to pursue the exploitation. Among the first to use the concepts of internal colonialism and institutionalized discrimination for the U.S. situation in the 1960s were Carmichael and Hamilton in *Black Power* (1967). They accented what they termed *institutional racism,* the actions taken by the white community as a whole against blacks as a group. Blacks are a colony in relation to white America in regard to housing, education, economics, and politics. Racial discrimination is thoroughly institutionalized and has persisted intact over a long period of time.

The historical dimension emphasized by Carmichael and Hamilton is particularly important, not the least because the class-over-race perspective plays down racial history. The U.S. city is not simply a social phenomenon overlaid with racism; it grew up in lockstep with racism. To understand modern racial issues in cities requires an understanding of historical roots. Briefly, the city has been racial from the beginning because the American colonies were grounded in a system of African and Afro-American labor exploitation. The first towns and cities of the colonies, and the great cities of the new nation after 1789, reflected from the inside out the realities of a brutal and complex system of Afro-American exploitation and subordination.

Becoming very prominent by the late 1700s, the forced labor in the port cities of the colonies, north and south, and in the fields of the southern plantations was an obvious aspect of the new nation's life. From the earliest years racial exploitation was coded into most basic institutions of the new nation, including its rural and urban economies, its political system, its educational system, and its urban housing patterns. The great cities of the new nation, from Boston, to Charleston, to New Orleans reflected these realities. Whether slave or free, Afro-Americans were relegated to the worst jobs and housing in every town and city of the colonies and, later, of the new nation. In the North, Afro-Americans were at first enslaved just as in the South. Slavery was, until well into the 1800s, seen by the majority of whites *in the North* as legitimate. The North was built in part on forced Afro-American labor and, as Ringer (1983, p. 533) puts it, "despite the [relatively] early emancipation of slaves in the North it remained there, not merely as fossilized remains but as a deeply engrained coding for the future." Coding here refers to the deeply imbedded system of laws and informal rules that have kept U.S. cities and U.S. society racist and segregated.

GUIDING CONCEPTS

Inspired by Carmichael and Hamilton's (1967) concept of institutional racism, Feagin and Feagin (1986) have defined racial and ethnic discrimination precisely as the *actions or practices carried out by members of dominant (racial or ethnic) groups that have a differential and negative impact on members of subordinate (racial and ethnic) groups*. The dimensions of this discrimination include the larger societal context, the immediate institutional context, the motivation (for example, prejudice) for the discrimination, the concrete discriminatory acts, and the effects of the acts (for example, resource inequality). Discriminatory actions against black and other non-European Americans take different forms in U.S. society; the forms include both individual and institutional (organizational) forms. An example of *individual discrimination* is the white homeowner who vandalizes a black family's home to drive that family out of a white neighborhood. This type of individual discrimination is common in the United States. Sometimes there are small-group conspiracies, such as those of violence-oriented white individuals who bomb a black family's home in a white neighborhood.

Institutionalized discrimination refers to organizationally prescribed actions carried out routinely by numerous white people against blacks in various organizational settings, such as manufacturing corporations and real estate businesses. The key feature is the way in which white people of differing persuasions carry out the discriminatory practices because they are expected in such contexts and are guided by informal norms and expectations. Examples of this type are also common; they include real estate agents' practices that routinely steer black families looking for housing away from white residential areas to predominantly black areas, and redlining practices by banks that make it difficult for black families to get housing loans in certain areas of cities. These practices perpetuate and reflect the ingrained racial coding that dates from slavery in the 17th century.

SIGNS OF INSTITUTIONALIZED RACISM: RESIDENTIAL SEGREGATION

MEASURING THE DEGREE OF GEOGRAPHICAL SEGREGATION

Residential segregation, one measure of the effects of institutionalized discrimination, is as old as the American city. While African American slaves may have been allowed to live in the back of white residences or on the back alley, free blacks have long been segregated in northern and southern cities, since before 1789. Most new groups entering the country have faced some housing discrimination and thus have faced some involuntary segregation. Research on the contemporary urban scene indicates, however, that segregation is greatest for black Americans. Taeuber's (1983) study of 28 central cities in larger metropolitan areas found only modest declines in black-white residential segregation between 1970 and 1980 (see Table 4.1).

Segregation decreases for most ethnic minorities with additional education, income, and occupational status (see Denton & Massey, 1988). This is not the case for Americans of African descent. The various measures of black socioeconomic status are not strongly correlated with the level of black segregation across urban areas (Darden, 1989; Massey & Denton, 1987). Jaynes and Williams (1989, pp. 144-146), in *A Common Destiny,* provide solid evidence that blacks of every class level are segregated from whites of similar status (see Table 4.2). Using the black-white residential segregation scores of 16 metropolitan areas, Jaynes and Williams (1989, p. 144) show that the segregation index for black families who had incomes of $50,000 and above equaled that of black families in poverty. Similarly, highly educated blacks are nearly as segregated as black high school dropouts.

When it comes to housing, the black middle class is confronted with the same harsh realities of racial discrimination and exclusion as its working class and underclass counterparts. Contrary to the "declining significance of

TABLE 4.1

U.S. Cities With Black Population of More Than 100,000 in 1980[a]

City	Population (in 1,000s)		% Black	Segregation Index (Dissimilarity)	
	Total	Black		1970	1980
New York, NY	7,071	1,784	25	77	75
Chicago, IL	3,003	1,197	40	93	92
Detroit, MI	1,203	759	63	82	73
Philadelphia, PA	1,688	639	38	84	88
Los Angeles, CA	2,967	505	17	90	81
Washington, DC	638	448	70	79	79
Houston, TX	1,594	440	28	93	81
Baltimore, MD	787	431	55	89	86
New Orleans, LA	557	308	55	84	96
Memphis, TN	646	308	48	92	85
Atlanta, GA	425	283	67	92	86
Dallas, TX	904	266	29	96	83
Cleveland, OH	574	251	44	90	91
Louis, MO	453	206	46	90	90
Newark, NJ	392	192	58	76	76
Oakland, CA	339	159	47	70	59
Birmingham, AL	284	158	56	92	85
Indianapolis, IN	701	153	22	90	83
Milwaukee, WI	636	147	23	88	80
Jacksonville, FL	541	137	25	94	82
Cincinnati, OH	385	130	34	84	79
Boston, MA	563	126	22	84	80
Columbus, OH	565	125	22	86	75
Kansas City, MO	448	123	27	90	86
Richmond, VA	219	112	51	91	79
Gary, IN	152	106	71	84	68
Nashville, TN[b]	456	108	23	90	80
Pittsburgh, PA	424	102	24	86	83

a. Segregation measures are based on the census count of black and nonblack persons in each city block.
b. Census data for Nashville in 1980 include all of Davidson County.
SOURCE: Taeuber (1983, Table 1).

race" argument, "black disadvantage resulting from housing segregation and or from white gain-motivated job discrimination is a result of racial processes, not class" (Farley, 1987, p. 147). Bart Landry, in *The Black Middle Class* (1987), offers some insights into the race-class debate and black housing patterns. Landry concludes that the "idea of a black middle class living in social isolation from other classes is largely a myth" (p. 185). At the

TABLE 4.2

Indices of Segregation by Income and Educational Attainment Level 1980

Income or Educational Level	Black-White Segregation in 16 Areas[a]
Family income in 1979	
Under $5,000	76
$5,000-$7,499	76
$7,500-$9,999	76
$10,000-$14,999	75
$15,000-$19,999	75
$20,000-$24,999	76
$25,000-$34,999	76
$35,000-$49,999	76
$50,000 or more	79
Educational attainment of persons 25 and over	
Less than 9 years	76
High school, 1-3 years	77
High school, 4 years	76
College, 1-3 years	74
College, 4 years or more	71

a. These residential segregation scores are average values for 16 metropolitan areas (Atlanta, Baltimore, Chicago, Cleveland, Dallas, Detroit, Houston, Los Angeles, Miami, New Orleans, New York, Newark, Philadelphia, St. Louis, San Francisco, and Washington, D.C.) computed from census tract data. The index shown for an income of $20,000-$24,999, 76, compared the residential distribution of black families in this income category to that of white families in the identical category.
SOURCE: Jaynes and Williams (1989, Tables 3-7).

rate the country is going, it will take more than six decades for blacks to achieve even the minimal levels of integration with whites that Asians and Hispanics have now.

THE DISCRIMINATION BEHIND METROPOLITAN SEGREGATION: GOVERNMENTAL PROGRAMS

Individual and institutionalized discrimination lie behind these statistical patterns of urban segregation. For two centuries federal government policies have played a key role in the development of spatially differentiated metropolitan areas where blacks and other visible minorities are segregated from whites and the poor from the more affluent citizens. This has been especially so since World War II (Goering, 1986; Momeni, 1986; Taeuber, 1983;

Taeuber & Taeuber, 1965; Tobin, 1987). Since then a basic dynamic creating racial polarization in U.S. cities has been the movement of white middle-income families to the suburbs, leaving behind a substantially poorer, often minority, population in the central cities. This suburban migration of whites has been intentionally stimulated by the investment decisions of white-run industrial corporations, banks, and developers and has been assisted by federal government subsidies for home mortgages and road building. In this uneven development process, capital flowed to housing in the suburbs and away from housing development in the central cities. Moreover, federal mortgage subsidies facilitated white movement out of the cities, at the same time that discriminatory federal restrictions (especially Federal Housing Administration rules) based on racist stereotypes about blacks and housing prohibited lending to most blacks desiring to move to the suburbs. Such intentionally racist policies fueled the white exodus to the suburbs and accelerated the abandonment of central cities (Kushner, 1980, p. 130). Federal tax dollars also funded the construction of freeway and interstate highway systems.

Many of the freeway construction projects intentionally cut paths through minority neighborhoods, physically isolated residents from their institutions, and disrupted once stable communities. In U.S. cities people of color are regularly displaced for highways, convention centers, sports arenas, and a host of downtown development projects. They are forced into other segregated areas with little input into the removal process (Feagin & Parker, 1990). The nation's apartheid-type policies have meant community displacement, gentrification, limited mobility, reduced housing options and residential packages, decreased environmental choices, and diminished job opportunities for the households who live in cities, while good jobs often move to the suburbs (Darden, 1989; Farley, 1987; Massey, Condran, & Denton, 1987; Massey & Eggers, 1990).

In a recent phase of the urban development cycle, investors have poured capital into certain central city areas, into the construction of new housing projects renovated to accommodate white professional and managerial families returning to central cities. Streets, schools, and parks were allowed to deteriorate. Now the whites moving back demand a change. The land occupied by blacks becomes a prime target for real estate development. The disinvestment-investment process has a racially discriminatory impact. This central-city seesawing may be rational from the points of view of profit-seeking white developers and the affluent white homeowners they serve, but it is clearly irrational for the minority urbanites whose lives are uprooted. New York's Harlem, originally constructed for middle-income whites but inhabited almost entirely by blacks between the 1920s and the 1980s, is one locality where real estate speculators have become conspicuous in the last decade. In one five-year period the number of building sales doubled annually, from 503 in 1980 to 1,131 in 1985. During the same period the average price per building rose from $130,000 to $230,000. During the late 1980s there were more daily property transfers in Harlem than elsewhere in Manhattan. Existing residents are less important than the future use, and probable destruction, of the buildings (Freedman, 1986, pp. 1, 19). Government has again played a role in this racial displacement. Speculative activity in Harlem has been fueled by the release of city-controlled (for example, formerly abandoned) housing. The New York City government owns about two thirds of Harlem's housing stock and, after curtailing its program of selling the tax-delinquent properties for several years in the 1980s, resumed public auctions of Harlem real estate in the 1980s. Harlem's city-aided speculative and redevelopment processes displace its long-term, usually black, residents, who are as a rule replaced by whites.

INSTITUTIONALIZED DISCRIMINATION: REAL ESTATE AGENTS AND LANDLORDS

Over five decades of federal housing policies, programs, and legislation have more often than not perpetuated racial barriers. And civil right legislation has so far had little effect. Title VI of the Civil Rights Act of 1964 and Title VIII of the Civil Rights Act of 1968 (Fair Housing Act) are two major pieces of federal legislation designed to remove the barriers to free choice in the housing market. The 1968 Fair Housing Act was a "paper tiger." The federal Fair Housing Act was amended and strengthened in 1988 because of the persisting problem of discrimination. Some federal government sponsored research in the late 1970s uncovered an alarming level of housing discrimination directed toward blacks (U.S. Department of Housing and Urban Development [HUD], 1979). The HUD study of 40 large metropolitan areas found that blacks faced a 72% chance of experiencing discrimination in the rental market and a 48% chance in the sales market. Racial "steering" (e.g., the practice of showing housing to one group only in certain areas) was rampant. The government report concluded that more than 70% of the whites and blacks who sought rental housing and 90% of those who were buying were steered into separate neighborhoods in the 1970s. Other studies, using fair housing audits, have documented the persistence of housing discrimination since that time. James, McCummings, and Tynan's (1984) study of the Houston, Denver, and Phoenix metropolitan areas, and Feins and Bratt's (1983) study of the Boston area document widespread discriminatory practices used by real estate managers to limit black and Hispanic housing choices.

The best studies have used a black auditor and a white auditor of similar backgrounds, who are sent to real estate agents and apartment agents. Studies in Dallas, Boston, and Denver found differential treatment favoring the white auditors. In all studies whites were more likely to be shown or told about more housing units than blacks, whites received more information on financing than blacks, and black housing seekers were steered away from traditionally white residential areas. Housing and employment discrimination directed against minority Americans is common in central cities and suburbs. There have been a number of important research studies that sent a trained black auditor and a white auditor (of similar socioeconomic backgrounds) to real estate agents selling homes and to apartment rental agents. In a 1980s Boston study the white auditors were invited to inspect 17 units on the average, 81% more than their black (matched) teammates. Another Boston study found a similar pattern of discriminatory treatment for blacks, Hispanics, and Asian Americans. And a 1980s Denver study found extensive discrimination against Hispanic owners and black renters across the city, while Hispanic renters and black owners faced high levels of discrimination in certain areas of the city. These studies have found that some real estate agents reserve some housing units to show to whites, and others for minorities. Whites are more likely to be encouraged to seek housing in suburban areas. A common pattern is for minorities to be shown advertised housing units, but not to be shown or told about other housing units that are available to whites. Highly segregated cities persist today two decades after the 1968 Civil Rights Act banning discrimination in housing (Yinger, 1986).

The 1988 Fair Housing Act is now several years old. It is too early to tell what impact it is having on housing barriers. But lawsuits under the act have revealed persisting discrimination by landlords. One of the first class-action lawsuits filed under this law was against an Orange County, California, firm that manages more than 4,000 apartment units in Los Angeles, Riverside, San Bernardino, and Orange counties. The 1989 lawsuit charged the company with using a "happy face" affixed to rental applications of minority home seekers for the purpose of screening out these applicants (Feldman, 1989). This well-institutionalized discriminatory practice is reminiscent of the "color codes" used by real estate agents to designate neighborhoods that were open and closed to blacks (Taub, Garth, & Dunham, 1984). One of the largest housing discrimination settlements in the country came in February 1990. It was made possible by the 1988 legislation. A Los Angeles rental complex (Belford Park Apartments) agreed to pay $450,000 to settle a lawsuit that demonstrated that apartment *owners* and *managers* engaged in pattern-and-practice of discrimination against black persons. The lawsuit, *Westside Fair Housing Council v. Westchester Investment Company,* had been litigated by the NAACP Legal Defense and Education Fund for more than two years. In addition to the monetary award, the consent decree called for the owner, Westchester Investment Company, to take affirmative steps to ensure that at least one fourth of the Belford Park's units have at least one black resident within three years (Prentice Hall, 1990, p. 2).

INSTITUTIONAL DISCRIMINATION: FINANCIAL INSTITUTIONS

The web of institutionalized discrimination in the housing market is a result of action and inaction—not only of legions of local and federal government officials and real estate marketing firms but also thousands of financial institutions and insurance companies (Jaynes & Williams, 1989; Yinger, 1986, pp. 140-144). The number of minority homeowners would be higher in the absence of discrimination by lending institutions (Darden, 1989). Only 59% of the nation's middle-class blacks own their home, compared with 74% of whites. Many inner-city neighborhoods have been strangled by the lack of long-term financing as a direct result of redlining practices by banks, savings and loans, mortgage firms, and insurance companies (Taggart & Smith, 1981). Moreover, banks and other lenders commonly set limits on real estate development by requiring that certain conditions be met before they will lend; financial institutions are in a position to dictate such conditions as size and scope of projects, type of financial statements, interest rates, bank access to records, legal fees, and even access to mass transit. In the case of smaller developers, finance capitalists can dictate virtually all important conditions. In making loans, lenders have considerable power to shape how, and whether, urban communities grow. When they have decided to finance corporations building office towers in central cities, they have contributed to the growth of administrative centers. When they have decided to deny loans to black homeowners in ghetto areas (a practice known as *redlining*), they have hastened the decline of housing in many low- to moderate-income areas. Lenders tend to prefer developers and industrial corporations as borrowers, because they figure a big company is less likely to go bankrupt than a small one. In this way they contribute to concentration and centralization in the real estate development sector of the U.S. economy.

The federal government recognized this problem when it passed the Community Reinvestment Act (CRA), a 1977 law designed to combat discriminatory practices in poor and minority neighborhoods. The CRA requires banks and thrifts to lend within the areas where their depositors live. The CRA has been used in conjunction with the Home Mortgage Disclosure Act, a law that requires banks and thrifts to disclose their mortgage lending by census tracts (Foust, 1987; Yang Oneal, & Anderson, 1988). Yet lending institutions persist in these redlining practices in spite of this type of federal initiative. Many of the same banks and thrifts that are now in default actively redlined minority and poor neighborhoods. Yet, all tax payers, including victims in the redlined neighborhoods, are being asked to bail out these often corrupt financial institutions.

OTHER WHITE INDIVIDUALS

Individual discrimination has also taken the form of violence. In one year (1985-1986) there were 45 known arson and cross burning attempts at the homes of blacks and other minorities who had moved into mostly white residential areas across the United States. In addition, during the 1980s there were hundreds of acts of vandalism and intimidation directed at black Americans living or traveling in white neighborhoods. One of the most famous acts of intimidation was in 1986 in the Howard Beach area of New York City where three blacks were beaten and chased by white youths. One black man died when he was chased into the path of a car (Feagin, 1990b).

SIGNS OF INSTITUTIONALIZED RACISM: AMENITY AND ENVIRONMENTAL DISCRIMINATION

DISCRIMINATION IN ZONING

All communities are not created equal. Race interpenetrates with the other factors and continues to be the most important variable in explaining the socio-spatial layout of urban areas, including housing patterns, street and highway configuration, commercial development, and industrial facility siting (Bullard, 1987, 1990; Feagin, 1988; Logan & Molotch, 1987). To this point we have focused on individual and institutionalized housing discrimination, including the street-level territoriality maintained by individuals. But there are other forms of residentially based discrimination as well. We can now turn to the related discrimination in public facilities and amenities. The differential residential amenities and land uses assigned in urban areas cannot be explained by socioeconomic status alone. Moreover, poor whites and poor blacks do not have the same opportunities to "vote with their feet" and escape undesirable physical environments. Minority and low-income residential areas (and their inhabitants) are often adversely affected by unregulated growth, ineffective regulations of industrial toxins, and public policy decisions authorizing locally unwanted land uses (LULUs) that favor those with political and economic clout (Bullard, 1983, 1990; Bullard & Wright, 1986, 1987).

Institutionalized discrimination can be seen in the widespread racial devices implemented by means of the zoning laws. Zoning is probably the most widely applied mechanism to regulate urban land use in the United States (see Babcock, 1966, pp. 3-18; Plotkin, 1987, pp. 75-110). Zoning laws broadly define land for residential, commercial, or industrial uses, and may impose narrow land-use restrictions (e.g., minimum and maximum lot sizes, number of dwellings per acre, square feet and height of buildings). Externalities such as pollution discharges to the air and water, noise, vibrations, and aesthetic disamenities are often segregated from residential areas because of the "public good." Nonresidential activities that are judged to have negative effects generally decrease with distance from the source. Zoning, thus, is designed as a "protectionist device" to insure a "place for everything and everything in its place" (Perrin, 1977). Zoning is ultimately intended to influence and shape urban land use in accordance with long-range local needs. Zoning, deed restrictions, and other land-use mechanisms have often been used as a "NIMBY" (not in my back yard) tool, operating through exclusionary practices. Exclusionary zoning, for example, means "simply to zone against something rather than for something" (Marshall, 1989, p. 312). Urban development and "spatial configuration" flow from the forces and relationship of production. These processes are dominated and subsidized by state actors (Feagin, 1988; Gottdiener, 1988). Implementation of zoning ordinances and land-use plans have a political, economic, and racial dimension. Competition often results between special interest groups (i.e., real estate interests, racial and ethnic minorities, organized civic clubs, neighborhood associations, environmentalists) for what these groups regard as more advantageous land use. The real estate interests generally have the most influence over zoning boards and local officials who make the decisions (Elkin, 1985; Logan & Molotch, 1987).

Governmental officials have done a poor job in protecting some communities, especially low-income, working-class, and black communities, against the ravages of industrial encroachment and environmental degradation. In their quest for quality neighborhoods, individuals often find themselves competing for desirable neighborhood amenities (i.e., good schools, police and fire protection, quality health care, and parks and recreational facilities) and resisting outputs that are viewed as having negative consequences (i.e., landfills, polluting industries, freeways, public housing projects, drug treatment facilities, and halfway houses). Black and other minority neighborhoods are usually the losers in this process because of the power of whites to control the allocation of urban disamenities using the zoning and deed restriction process.

USING DEED RESTRICTIONS
AND GOVERNMENT ACTORS

Even in cities without zoning, minority neighborhoods encounter the institutionalized racism of disamenity locations. Deed restrictions and developer decisions are the mechanisms of intentional racial discrimination. Thus Houston, the nation's fourth largest city and only major American city without zoning, is a classic example where "unrestrained capitalism" has dominated the spatial configuration, built environment, and land-use outcomes (Bullard, 1987; Feagin, 1988). The city's landscape has been shaped by haphazard and irrational land-use planning, a pattern characterized by excessive infrastructure chaos (see Babcock, 1982; Bullard, 1983; Feagin, 1988). In the absence of zoning, developers have used renewable deed restrictions and pressures on white government officials to control land use with subdivisions. Deed restrictions work, however, only for the white affluent communities with the funds to hire lawyers to enforce them against encroachment. They are ineffective land-use controls, especially in black and other minority neighborhoods. Weakly enforced or nonexistent regulations have created a nightmare for many of these black neighborhoods, areas that are ill-equipped to fend off industrial and waste dump encroachment. Black Houston, for example, has had to contend with a disproportionately large share of the city's garbage dumps, landfills, incinerators, salvage yards, and a host of other LULUs. This has historically been a blatant type of well-institutionalized discrimination perpetuated by several generations of white developers, political officials, and neighborhood organizations. The siting of nonresidential facilities, as in the case of solid waste sites, has heightened animosities between Houston's black community and the white-controlled city government. The burden of having a municipal landfill or incinerator near one's home has not been equally shared by all Houstonians. Black Houston has become the dumping ground for the city's white and minority household garbage (Bullard, 1983).

Houston operated five landfills, from the 1920s to the early 1970s, to dispose of its garbage. All five of the city-owned landfills were located in black neighborhoods. Houston operated eight incinerators during this same period. Six of the eight city-owned garbage incinerators were located in black neighborhoods. The city closed its waste facilities in the early 1970s and contracted out its waste disposal services with private firms. From the early 1970s to the late 1970s, four privately owned landfills were used to dispose of Houston's solid waste. Three of these facilities were located in mostly black neighborhoods, although blacks made up just one fourth of the city's population. The private disposal industry followed the discriminatory siting pattern that had been established by the Houston city government.

But Houston is by no means alone in its patterns of disamenity discrimination. A *Toxic Wastes and Race* report (Commission for Racial Justice, 1987) was written to challenge this amenity and environmental racism. This national study found race to be the *single* most important factor—more important than income, home ownership rate, and property values—in the location of abandoned toxic waste sites across urban landscapes. Urban minority populations are exposed to greater risks from the nation's uncontrolled (abandoned) waste sites than their nonminority counterpart. For example, Atlanta has a total of 94 uncontrolled toxic waste sites. Nearly 83% of the city's black population live in zip codes where these sites are found, while 60% of white Atlantans live in similar waste site areas (Commission for Racial Justice, 1987). Although tagged the "capital of the New South," Atlanta remains one of the most racially segregated cities in the region.

Blacks are not the only victims of disamenity discrimination. A similar pattern is repeated in cities that have large concentrations of Hispanics. Los Angeles, the nation's second largest city, has a total of 60 uncontrolled toxic waste sites. Some 60% of the city's Hispanic population live in waste site areas compared with 35% of the city's whites. Similarly, Houston has 152 uncontrolled toxic waste sites. More than 81% of Houston's Hispanic population live in waste site areas compared with 57% of the city's whites. Individuals who are in close proximity to noxious facilities and pollution are living in endangered environs. The price they pay is in the form of higher risks of emphysema, chronic bronchitis, and other chronic pulmonary diseases. In addition, air pollution in inner-city neighborhoods can be found at levels up to five times greater than those found in suburban areas.

NEW WAVES OF ENVIRONMENTAL DISCRIMINATION

Urban black and other minority neighborhoods have become disamenity dumping grounds and environmental wastelands because of decades-old discrimination by white officials and developers to locate disamenities in minority neighborhoods. This is not a new type of institutionalized discrimination. But there has been a new wave of reinforcement of this discrimination. People of color, at every class level, are now the targets of industrial dumping that has skyrocketed over the last several decades of growth in our chemical-centered society (see Bullard, 1990; Commission for Racial Justice, 1987; U.S. General Accounting Office, 1983). This problem is repeated in every region of the country. The Reverend Benjamin Chavis of the Commission for Racial Justice equates toxic dumping in minority communities as tantamount to "environmental racism." For Reverend Chavis, environmental racism is "enforced and maintained by the legal, cultural, religious, educational, economic, political, environmental and military institutions" (Commission for Racial Justice, 1987, p. x). Moreover, the white-middle-class-dominated environmental movement of the 1960s and 1970s built an impressive political base for environmental reform and regulatory relief from the disamenities of the chemical-centered industrial society. Most white environmentalists, however, gave little attention to the sociological implications of the NIMBY phenomenon (Morrison, 1986). Given the political climate of the times, the hazardous-waste facilities, garbage dumps, and polluting industries were likely to end up in somebody's backyard. But whose backyard? Very often these LULUs ended up in powerless minority communities rather than in affluent white suburbs (Bullard, 1990). This pattern has proven to be the rule, even though the benefits derived from industrial production are directly related to affluence. Public officials and private industry have in many cases responded to the NIMBY phenomenon using the "PIBBY" principle, "Place in Blacks Back Yard" (Bullard & Wright, 1987, p. 28).

CONCLUSION

Most "race and the city" discussions today focus on issues of poverty and the underclass. And much of the theoretical underpinning for this discussion is economic, a type of market-centered economics. Racial issues are reduced to economic issues. But race must be treated as an independent variable. We have seen that the modern American city has its roots in well-institutionalized racism. This racism can be seen in its basic ecological form. Racial segregation in housing, as well as schools and jobs, is fundamental to the geography of the modern city. Many social scientists for decades have described the racial and ethnic segregation of cities and the slow integration of ethnic groups into the host neighborhoods. Yet the oldest of the American minority immigrant groups, black Americans, has been the least integrated into white institutional arenas. This pattern is fundamentally rooted in the color coding arising out of and essential to slavery and subsequent institutionalized discrimination and segregation.

The development, housing, and environmental discrimination described in this chapter are not the result of some impersonal superstructural process. Many urban ecologists have accented what they see as an impersonal segregation process whereby groups stay with their own kind, and an impersonal invasion-succession process whereby one race/ethnic group more or less inevitably replaces another—thereby making up the urban racial-ethnic mosaic. The racial patterns of cities have not been created by some remote or impersonal process. Specific actors "did it"; and the actors were quite an array of whites: the white slaveholders, merchants, and shippers of the early period and the white landlords, real estate agents, business elites, and politicians in the periods since slavery.

If change is to come in the racist American city, this change will also require the actions of these white actors, who may, one can hope, be made to see that the survival of the U.S. city and the United States in a world populated mostly by people of color is conditional on a destruction of racism at home. Black Americans bear the brunt of continuing institutionalized racism, but urban racial inequalities have enormous social and financial costs for the white population as well. Billy Tidwell of the National Urban League estimates that discrimination against blacks lowers the nation's gross national product by almost 2% a year, or roughly $104 billion in 1989 (Updegrade, 1989). A large share of this loss is a result of housing discrimination. The separation of blacks in central cities and whites in suburbs has other serious implications as well. There is of course the problem of blacks' getting jobs developed in suburban areas. But here is also a major problem for suburban whites. The future white leaders who grow up in suburban enclaves will have no ability to talk to and relate to peoples of color. In a country expected to be more than one-third minority in its population composition over the next few decades this is a very serious problem for whites. We can note again that the Kerner commission report on the black city riots of the 1960s warned that the United States was then resegregating itself into two "separate and unequal" societies. This remains a fundamental U.S. problem.

NOTE

1. See, the example, Nathan Glazer, *Affirmative Discrimination* (New York: Basic Books, 1975). Glazer has described the white-black problem in terms of a "tangle of pathology in the ghetto" and argued that neither rapid economic growth nor affirmative action would benefit

those impoverished blacks; Glazer was among the first to argue that affirmative action programs had major discriminatory effects on whites.

REFERENCES

Babcock, R. F. (1966). *The zoning game.* Madison: University of Wisconsin Press.

Babcock, R. F. (1982). Houston: Unzoned, and mostly unrepentent. *Planning* 48: 21-23.

Black and white in America. (1988, March 7). *Newsweek,* pp. 19-20.

Bowser, B. P., & Hunt, R. G. (1981). *Impact of racism on white Americans.* Beverly Hills, CA: Sage.

Bullard, R. D. (1983). Solid waste sites and the black Houston community. *Sociological Inquiry, 53,* 273-288.

Bullard, R. D. (1987). *Invisible Houston: The black experience in boom and bust.* College Station: Texas A & M University Press.

Bullard, R. D. (1990). *Dumping in Dixie: Race, class, and environmental quality.* Boulder, CO: Westview.

Bullard, R. D., & Wright, B. H. (1986). The politics of pollution: Implications for the black community. *Phylon, 47,* 71-78.

Bullard, R. D., & Wright, B. H. (1987). Environmentalism and the politics of equity: Emergent trend in the black community. *Mid-American Review of Sociology, 12,* 21-37.

Carmichael, S., & Hamilton, C. (1967). *Black Power.* New York: Vintage.

Commission for Racial Justice. (1987). *Toxic wastes and race: A national report on the racial and socioeconomic characteristics of communities with hazardous wastes sites.* New York: United Church of Christ Commission for Racial Justice.

Darden, J. T. (1989). The status of urban blacks 25 years after the civil rights act of 1964. *Sociology and Social Research, 73,* 160-173.

Denton, N. A., & Massey, D. S. (1988). Residential segregation of blacks, Hispanics, and Asians by socioeconomic status and generation. *Social Science Quarterly, 69,* 797-817.

Elkin, S. L. (1985). Twentieth century urban regimes. *Journal of Urban Affairs, 7,* 11-28.

Farley, J. E. (1987). Disproportionate black and Hispanic unemployment in U.S. metropolitan areas: The roles of racial inequality, segregation and discrimination in male joblessness. *American Journal of Economics and Sociology, 46,* 129-150.

Feagin, J. R. (1988). *Free enterprise city: Houston in political and economic perspective.* New Brunswick, NJ: Rutgers University Press.

Feagin, J. R. (1990a). *Building American cities: The urban real estate game.* Englewood Cliffs: Prentice Hall.

Feagin, J. R. (1990b). *Racial and ethnic relations* (3rd ed.). Englewood Cliffs, NJ: Prentice-Hall.

Feagin, J. R., & Feagin, C. B. (1986). *Discrimination American style: Institutional racism and sexism.* Malabar, FL: Robert E. Krieger.

Feagin, J. R., & Parker, R. (1990). *Building American cities: The urban real estate game.* Englewood Cliffs, NJ: Prentice-Hall.

Feins, J. D., & Bratt, R. G. (1983). Barred in Boston: Racial discrimination in housing. *Journal of the American Planning Association, 49,* 344-355.

Feldman, P. (1989, October 26). Suit charged large apartment firm with racial bias in rentals. *Los Angeles Times.*

Foust, D. (1987, March 2). Leaning on banks to lend to the poor. *Business Week, 2987,* 76.

Freedman, S. G. (1986, December 19). Harlem and the speculators: Big profits but little renewal. *The New York Times,* pp. 1, 19.

Glazer, N. (1975). *Affirmative discrimination.* New York: Basic Books.

Goering, J. M. (1986). *Housing desegregation and federal policy.* Chapel Hill: University of North Carolina Press.

Gottdiener, M. (1988). *The social production of urban space.* Austin: University of Texas Press.

James, F. J., McCummings, B. I., & Tynan, E. A., (1984). *Minorities in the sunbelt.* New Brunswick, NJ: Rutgers University, Center for Urban Policy Research.

Jaynes, G. D., & Williams, R. M., Jr. (1989). *A common destiny: Blacks and American society.* Washington, DC: National Academy Press.

Kelly, E. D. (1988). "Zoning." In F.S. and So and J. Getzels (eds.), *The practice of local government planning* (2nd ed, pp. 251-284). Washington, DC: International City Management Association.

Kerner Commission. (1967). *Report of the national advisory commission on civil disorders.* New York: Bantam Books.

Kushner, J. A. (1980). *Apartheid in America: An historical and legal analysis of contemporary racial segregation in the United States.* Frederick, MD: Associated Faculty Press.

Landry, B. (1987). *The new black middle class.* Berkeley: University of California Press.

Logan, J. H., Molotch, H. (1987). *Urban fortunes: The political economy of place.* Berkeley: University of California Press.

Marshall, P. G. (1989, June 9). Not in my back yard. *Editorial Research Reports, 1* , 306-319.

Massey, D. S., Condran, G. A., & Denton, N. A. (1987). The effect of residential segregation on black social and economic well-being. *Social Forces, 66,* 29-56.

Massey, D. S., & Denton, N.A. (1987). Trends in segregation of blacks, Hispanics and Asians, 1970-1980. *American Sociological Review, 52,* 802-825.

Massey, D. S., & Eggers, M. L. (1990). The ecology of inequality: Minorities and the concentration of poverty, 1970-1980. *American Journal of Sociology, 95,* 1153-1188.

Momeni, J. A. (1986). *Race, ethnicity, and minority housing in the United States.* Westport, CT: Greenwood Press.

Morrison, D. E. (1986). Why and how environmental consciousness has trickled down. In A. Schnaiberg, N. Watts, & K. Zimmerman (Eds.), *Distributional conflict in environmental-resource policy* (pp. 187-220). *New York: St. Martin's.*

NAACP Legal Defense and Educational Fund. (1989). *The unfinished agenda and race in America.* New York: NAACP Legal Defense and Educational Fund.

National Advisory Commission on Civil Disorders. (1968). Report of the National Advisory Commission on Civil Disorders. Washington, D.C.: U.S. Government Printing Office.

Perrin, C. (1977). *Everything in its place: Social order and land use in America.* Princeton, NJ: Princeton University Press.

Plotkin, S. (1987). *Keep out: The struggle for land use control.* Berkeley: University of California Press.

Prentice Hall (1990, April 1). Rejection of black roommate brings $450,000 settlement in Los Angeles lawsuit. *Fair Housing-Fair Lending, 5,* pp. 1-3.

Ringer, B. (1983). *"We the people" and others.* New York: Tavistock.

Taeuber, K. (1983). *Racial residential segregation, 28 cities, 1970-1980* (Center for Demography and Ecology Working Paper 83-12). Madison: University of Wisconsin.

Taeuber, K., & Taeuber, A. K. (1965). *Negroes in cities: Racial segregation and neighborhood change.* Chicago: Aldine.

Taggart, H. T., & Smith, K. W. (1981). Redlining: An assessment of the evidence of disinvestment in metropolitan Boston. *Urban Affairs Quarterly, 17,* 91-107.

Taub, R. P., Garth, D., & Dunham, J. D. (1984). *Paths of neighborhood change: Race and crime in urban America.* Chicago: University of Chicago Press.

Tobin, G. A. (1987). *Divided neighborhoods: Changing patterns of racial segregation.* Newbury Park, CA: Sage.

Updegrade, W. L. (1989). Race and money. *Money, 18,* 152-172.

U.S. Department of Housing and Urban Development. (1979). *Measuring racial discrimination.* Washington, DC: Government Printing Office.

U.S. General Accounting Office. (1983). *Siting of hazardous waste landfills and their correlation with racial and economic status of surrounding communities.* Washington, DC: General Accounting Office.

Whites, blacks hold different views of status of blacks in U.S. (1980, June). *Gallup Opinion Index* (Report No. 178), p. 10.

Wieseltier, L. (1989, June 5). Scar tissue. *The New Republic,* 19-20.

Wilson, W. J. (1978). *The declining significance of race: Blacks and changing American institutions.* Chicago: University of Chicago Press.

Yang, C. M., Oneal, & Anderson, R. (1988, August 15). The "blackmail" making banks better neighbors. *Business Week,* 3065, p. 101

Yinger, J. (1986). Measuring racial discrimination with fair housing audits. *American Economic Review, 76,* 881-893.

5

Restructuring Production and Reproduction: Some Theoretical and Empirical Issues Relating to Gender, or Women in Britain

LINDA McDOWELL

ALTHOUGH THIS IS THE CHAPTER that the editors commissioned to provide an explicit feminist perspective, or at least to raise questions about the implications for women of the far-reaching social changes that are restructuring contemporary industrial societies, I want to suggest that it might be argued that the purpose, if not the execution, of the book as a whole is (implicitly) feminist. The collection arose from the editors' recognition that the economic restructuring currently transforming the west raises a series of questions that are inadequately conceptualized and dealt with in most of the theoretical debates. Different schools of thought—from the flexible accumulation theorists, through the deindustrialization school, to the postindustrialists' emphasis on technological change—focus almost exclusively on economic considerations, seldom expanding their analyses to include the concurrent changes in family life, social relations, and community involvement in a whole range of noneconomic struggles. Their recognition of the partiality of contemporary theories of crisis and restructuring thus parallels the long tradition of feminist critique of "malestream" social, economic, and political theory. For the focus on the economic in contemporary urban studies and geography reflects the separation of the public and the private and its associated sexual division: where the former is the masculine public work of competition, individualism, legal rights, and waged work. Theorists focus on this sphere in contrast with, and in opposition to, the private world of emotion, obligations, love, and subjectivity, a feminine sphere that is the "natural" foundation of civil life requiring no theorization. Thus contemporary theories of restructuring concentrate on the public world, occasionally addressing women's increased entry into the sphere of capitalist wage relations, but confining the private realm of the family and (to a lesser extent) the community to a separate arena, untouched by theoretical scrutiny. Relations of production are prioritized, those of consumption and reproduction generally ignored. But the contemporary restructuring of the economy raises important questions about the nature of the division between the public and the private, about the allocation of new jobs among categories of workers, about familial and household allocation of earnings, about noneconomic power relations and the gendered basis of these divisions.

While some theorists assert women's continued marginality to the economy and others see economic restructuring as challenging the basis of patriarchal relations, most analyses of contemporary change, other than explicitly feminist scholarship, continue to conceptualize the relations of production and reproduction separately, in isolation one from the other. It is in this continued separation that perhaps the current collection fails to become more than implicitly feminist. In this chapter I shall argue that it is essential to theorize and empirically investigate together changes in the arena of waged work and of all the unwaged labor that goes into reproducing workers and their dependents in order to grasp the causes and consequences of economic restructuring. The argument will be illustrated with British examples, although studies of women's paid and unpaid work in other industrial societies and in contemporary Third World countries also suggest that a satisfactory understanding of the nature of recent economic change requires the mutual analysis of production and reproduction, of class and gender relations.

CLASS AND GENDER RELATIONS IN THE "NEW ORDER": A BRIEF SUMMARY

Although contemporary industrial societies are distinguishable by the pace of economic and social change, in recent decades the transformations in the economy, in the nature of work, in social relations, in the structure of families and households appear to have been so significant that a theoretical debate about the nature of this "new order" has become dominant across a wide spectrum of academic disciplines and in policy circles. Changes in the international world order have reshaped the political agenda—the political positions of East and West Europe, of Britain, the United States and the U.S.S.R. have altered dramatically, the implications of the Gulf crisis for world trade in oil and the economic basis of the industrialized economies are enormous. Within these countries, the internal structure of the economy has been transformed by the shift from manufacturing to service-sector employment. In conjunction with technological change, the nature of work has been transformed for large sectors of the labor force and new divisions have opened up between highly skilled and well paid white-collar occupations and a growing number of part-time, casual, and poorly remunerated jobs at the bottom end of the occupational hierarchy. This latter group of workers may no longer experience lifelong attachment to the labor force in a single occupation that was the characteristic pattern of work, for most men at least, between the end of World War II and some time in the 1970s. Disrupted patterns of attachment to the labor market are an increased feature of a growing proportion of people's job histories and another growing group are excluded entirely, as levels of unemployment in the industrial West remain high for those with few skills or the social attributes valued in the new "postindustrial" economy.

Industrial restructuring and technical change have brought with them not only changes in employment patterns, and in the internal regional geographies of industrial societies, but also profound changes in patterns of income distribution, both among individuals and between and within households, changes in family living arrangements and in the position of women. Increasing rates of divorce, declining fertility rates, and increasing conceptions outside marriage are typical features of industrial western societies. Changes in individual social arrangements have been accompanied by a restructuring of collective welfare provisions in many of these countries. The 1980s have witnessed the ascendancy of the rhetoric of the "new right," with clarion calls for individuals to accept moral responsibility for their own lives, rejecting the dependency encouraged by the "nanny" state. The resultant cuts in collective provision of health care, housing, social income to support the low paid and unpaid have had a severe impact on the living standards of the poorest members of society, exacerbating the divisions opened up by economic restructuring. While theorists of the new order have not ignored the consequences of these far-reaching social changes, the debate has tended to focus on the consequences of industrial restructuring for the nature of class relations. A common theme has been the social and political effects of the so-called end of the working class as manufacturing jobs have reduced secure, relatively well-paid employment opportunities, previously predominantly for men (Gorz, 1982). But the new order has also brought with it changes in the nature of gender relations, in the position of women and men in advanced industrial nations, in the structure of power and social divisions between them. Women are entering waged labor in unprecedented numbers in these societies (and in many parts of the Third World), with concomitant effects on their standard of living, purchasing power, and responsibilities and rights within the family and in the household. In combination with changes in the structure of welfare provision, the net result has been a significant increase in the amount of labor performed by most women, but particularly by working-class women.

STRUCTURE OF THE CHAPTER

The overall purpose of this chapter is to address the question of how the restructuring of the economy and the social institutions of welfare provision matter to women, how structural changes are related to changes in individual living arrangements and in the structure of households and women's roles within them. In the first section, the extent to which current theories of economic restructuring allow a purchase on these issues will be examined. In the next section, through an empirical investigation of women's changing position in the British labor market, it is argued that the recent social and economic changes may be more adequately understood when gender relations are a central focus of the analysis. It is not sufficient to add women into existing theories, to append a question about the consequences for women. Rather gender must be seen as a constitutive part of the restructuring process. A decade or more of studies of the labor process by feminists have revealed that gender relations are centrally implicated in the organization of production, in the labor process and in the wage relation. Gender specific attributes are a key part of the notion of skill, of the differential relation of the sexes to technological change, to the definition of jobs as male or female and in their financial rewards. And as earlier feminist analyses of domestic labor demonstrated, the organization of what is traditionally regarded as the economy cannot be separated from the domestic and community organization of the "unproductive" work of daily and generational reproduction. Processes and relations in the economy and in society are fundamentally gendered. In the succeeding two sections of the chapter, social changes associated with economic restructuring are examined—first recent demographic change and then the social welfare policies of the Thatcher governments. In this second section it will be demonstrated that the patriarchal ideology and practices of the state in this area are raising a contradiction between women's role in the economy and in their unpaid caring and servicing work: a crisis that variants of the restructuring thesis are ill-placed to understand. However, the old predictions of socialist feminism that labor force participation by women was a precursor of liberation also stand in need of reevaluation. Indeed, the assumption that capitalism needs domestic labor that lies at the base of socialist-feminist theory is also being challenged by contemporary changes. This is briefly discussed in the penultimate section of the chapter.

VARIANTS OF RESTRUCTURING THEORY: FLEXIBLE SPECIALIZATION AND REGIMES OF ACCUMULATION

A number of separate, but related, versions of the nature of the transformation in the industrial economies may be identified. Despite the differences between the adherents of particular perspectives, a number of features unite the contributors to the debate. First, there is a common emphasis on the impact of deindustrialization, on the causes and consequences of the shift to a service-based economy. Second, and somewhat paradoxically, theorists of the new order share focus on transformations in the methods of industrial production, and hence place a continued emphasis on the manufacturing sector. Thus the regulation theorists have identified a new regime of accumulation that is distinguished from the former Fordist regime by new flexible forms of the organization of production (Aglietta, 1979; Lipietz, 1987; Storper & Scott, 1988; Storper & Walker, 1989). Technological and market changes in combination have overturned the conditions for Fordist assembly line methods of mass production on Taylorist principles. These are being replaced by functional flexibility, just-in-time ordering systems, multipurpose equipment, team working, and Japanese-style quality circles. Third, the theorists have in common an emphasis on the changing geography of production: the emergence of new regional economies or new industrial spaces characterized by a network of flexible firms in contradistinction to old patterns of concentration and dominance by a single large firm or sector in the old industrial heartlands of the Fordist era (Scott, 1988).

The evidence for these changes and the emergence of new regional economies is limited and depends in the main on studies of manufacturing industry, particularly the automobile, defense, and aircraft industries. While criticisms of the extent to which a new hegemonic order is emerging and of its geographical form are increasingly common (Amin, 1989; Amin & Robins, 1990; Lovering, 1990; Pollert, 1988; Sayer, 1990) critiques of the particular way in which gender relations, or rather women's role in economic change, have been dealt with in restructuring theories are less widespread.

Here two versions of the flexible accumulation models are examined, assessing the ways in which assumptions about gender divisions are explicitly or implicitly part of the analysis. The two approaches are the vision of the future exemplified in Piore and Sabel's work and the "new industrial geography" school of the West Coast U.S. theorists, the main proponents of which include Scott, Storper, and Walker.

GENDER, THE FLEXIBLE WORKER, AND SKILLS

Piore and Sabel's version of the future spelled out in their book *The Second Industrial Divide: Possibilities for Prosperity* (1984) is an essentially optimistic one, based in the main on the potentially liberating possibilities of technological change. They argue that the advanced industrial countries of the world are witnessing a shift away from mass production processes based on special purpose machinery in which semiskilled workers perform monotonous tasks to a new world of work in which highly skilled workers, equipped with technical know-how, perform multipurpose tasks, producing goods for rapidly changing markets, based on multiply differentiated tastes. Such changes are assumed to lead to new, less conflictual management-labor relations in which workers are able to exercise group control over their working conditions. Nowhere in their book, however, is there a recognition of the gendering of tasks that are defined as skilled. Skill is a socially constructed concept and as such the definition of particular jobs and tasks as skilled has been subject to struggle and renegotiation, not only between capitalists and workers, or between groups of male workers, but also between men and women. There is a well-documented social history of the struggles between craft-based organizations and the early industrial unions that challenged both the definitions of craft and the exclusionary practices based on these definitions. However, the parallel ways in which male workers organized to oppose women's entry into skilled occupations have not been as well-documented until the recent feminist insistence that the definition of skill embodies gender-specific attributes (Phillips & Taylor, 1980).

There is now a growing body of literature about the relationship among new technology, skill, and gender that extends our understanding of the ways in which gender relations are reproduced in work relations. Cockburn's pioneering work on the printing industry (1983) and on other occupations in which the introduction of new technology resulted in the redefinition of skills (Cockburn, 1985) cast doubt on the emancipatory possibilities of technical and technological change for women. As Cockburn and others (see for example, Game & Pringle's [1984] work in Australia) have shown, jobs themselves are gendered, they exist and change in social relations of hierarchy that reproduce unequal relations of power for men and for women. "The hierarchies of skill are not just an imperative of capitalist production, they express at the same time a system of male dominance . . . jobs are created as masculine and feminine, with their skill content continually redrawn to assert the dominance of men" (Phillips, 1983, p. 102). In particular, those jobs that entail control over the manipulation of technology are designated as "male" jobs. The jobs that women do either do not entail such manipulation or the tasks that women do are regarded as unskilled because they draw on "natural" feminine attributes, particularly those present in domestic tasks. These assumptions are at the heart of recent challenges to gender segregation in the labor market embodied in the campaigns for equal pay for work of equal value.[1]

The prevailing societal definition of femininity is based on the idea that familiarity with machinery is somehow unfeminine or de-sexing for women. This ideology pervades the design and construction of machinery itself—the "average" worker for whom it is designed is male—as well as the allocation of tasks by management and women's own identity and relations at work. Women themselves are often reluctant to acquire technical skills, which are seen as unfeminine (Cockburn, 1986), and in addition may regard their waged work as relatively unimportant, escaping the monotony of unskilled employment by recreating familial situations at work or by escaping into daydreams of romance and marriage (Westwood, 1984). For all these reasons the skilled workers on whom Piore and Sabel (1984) pin their version of the future are male.[2] Restructuring the labor process to privilege skilled work and workers thus recreates the gender segmentation of earlier methods of industrial production.

To the extent that women do enter the analysis, they are there in their familiar guise as "marginal" workers, participating in the secondary labor market along with other "minority" groups. Thus Piore, writing earlier with Berger (Berger & Piore, 1980), suggests that

> the migrants (foreign and domestic), the rural workers and the women (*how dismissive is that definite article*) are attractive precisely because they belong to another socio-economic structure and view industrial employment as a secondary adjunct to their primary roles. They are willing to take jobs because they see their commitment to these jobs as temporary, and they are able to bear the flux and uncertainty of the industrial economy because they have traditional activities on which to fall back. (p. 50)

This view of women as secondary workers, dependent on men and with "traditional activities" (presumably housework) to fall back on not only denies the contemporary reality of manufacturing decline and the loss of the "family" wage for the male labor aristocracy but also changing familial structures and the growth of female-headed households (see below). In addition, it is an incorrect reading of women's labor market participation patterns. As Walby (1989), among others, has demonstrated, the notion of women as a flexible reserve, drawn into and expelled from the economy at times of expansion and recession, is incorrect. Women's paid work in Great Britain— –both full-time and part-time—has continued to expand throughout the entire postwar period. Even during the depths of the recession in the early 1980s, when the absolute increase was temporarily halted, women's employment expanded relative to men's.

WOMEN'S WORK AND REGIMES OF ACCUMULATION: THE "NEW INDUSTRIAL GEOGRAPHERS"

The unproblematized allocation of women to the peripheral, casualized labor market is also a feature of the work of the "new industrial geographers." Hence Storper and Scott's (1988) distinction between "a highly remunerated segment consisting of professional, craft, and technical workers, and a poorly remunerated segment made up of politically marginalized social groups such as women, ethnic minorities and rural-urban migrants" (p. 32). The same groups are allocated to the periphery, although this time at least it is "women," not the condescending "the women" and the insertion of the adjective "political" before the term marginal seems to be at least a step toward recognizing that there are a set of *social* mechanisms at work behind this secondary position, other than women's role in the family. The particular ways in which gender-specific attributes are implicated in women's exclusion from highly remunerated jobs remain unexplored.

The work by the "new" geographers, however, with its origins in the French regimes of regulation school[3] (Aglietta, 1979) does partially take on board the feminist insistence of the necessity of analyzing the interconnection of the spheres of production and reproduction in understanding contemporary socioeconomic change—albeit unsatisfactorily. Storper and Scott (1988) define the new flexible production agglomerations as "an interdependent system of production spaces (the locus of work) [actually paid work] and adjunct social spaces (the locus of domestic life of the worker) [only an adjunct!—and where have we come across that term *adjunct* before? and what about the growing proportion of the population with no waged work?]"

(p. 33, comments added). They extend this definition of adjunct space to include

> distinctive neighborhoods and communities within which subtle and intricate processes of family life, childrearing and social interaction take place. These social areas then function as sites of symbolic representation of social distinction, identity and status, as portrayed by Baudrillard (1973) and Bourdieu (1979) with their theories of the political semiology of consumption and cultural activity. In short, the distinctive neighborhoods and communities that emerge within any agglomeration become integral to the legitimation and stabilization of socio-economic divisions in the local area. (pp. 33-34)

This statement contains a number of problems. Despite their recognition of the interdependence of production and reproduction, the significance of the "adjunct social spheres" *and* of production spaces as the locus in which gender divisions are created and recreated is ignored. It is now recognized that the construction of the gendered subject is a continuous process that occurs not only in the arena of civil society but also in the workplace. Attributes of masculinity and femininity are not only inscribed within occupations but also constructed and reconstructed in daily interaction in the labor market. Second, the home and the community are not only the spheres of consumption and cultural activity, but also of unwaged labor that is essential to the continuance and the social relations of waged labor. The reliance of Baudrillard and Bourieu, who also ignore the significance of gender divisions, is unhelpful here. Finally, by confining any consideration to the sphere of reproduction to the *local* area, the larger scale significance of the organization of unwaged labor and the restructuring of the institutions of collective welfare provision at the national level and their relationship to economic restructuring is missed. This is despite the argument of the regulation theorists that each regime of accumulation is associated with a particular mode of social regulation and institutions of social reproduction at the national level.

WOMEN'S WORK IN GREAT BRITAIN

In the next two sections, the ways in which restructuring have affected women's waged and unwaged work in Great Britain will be examined. I shall demonstrate that, viewed from the perspective of women, the "new" transformation seems not to be that new at all. Rather it is based upon the

TABLE 5.1

Trends in Men's and Women's Employment—Great Britain:
Employees in Employment (June)(in thousands) 1971-1988

	Men	*Women*	*Women as % of total*	*Women full-timers as % of total*	*Women part-timers as % of total*	*Women part-timers as % of all women*
1971	13,424	8,224	38.0	25.3	12.6	33.5
1976	13,097	8,951	40.4	24.3	16.3	39.6
1981	12,278	9,108	42.6	24.7	17.8	41.9
1986	11,643	9,462	44.3	25.2	19.6	42.4
1988	11,979	10,096	45.7	26.2	19.5	42.8

SOURCE: *Department of Employment Gazette,* various dates.

continuation of old forms of gender division and male dominance. Recent technological change and the transformation of work do not appear to be breaking down conventional gender stereotypes or the gender segmentation of the labor force, and the retreat from collective forms of social provision is increasing the burden of women's unwaged reproductive labor. In short, the transformation distinguished by many theorists is reinforced, indeed under-pinned, by the unchanging nature of long-established patterns of gender segregation. This is not to deny, of course, that significant changes have occurred in the ways in which women organize their lives, live with or without men, in different family and household forms. Women in Britain probably have greater access to economic resources, particularly from wages, in comparison with men that they have had before and many women enjoy a far wider range of social and cultural opportunities than was ever envisaged possible or appropriate in previous periods. But not withstanding these changes, gender differentiation and women's oppression by men remains a structural feature of British economic, social, and political institutions and practices.

WOMEN IN THE LABOR MARKET: SEX SEGREGATION RULES OK?

Women's entry into waged work has been on a previously unprecedented scale in Britain in recent decades. Labor force participation rates by women in the economically active age group (16 [15 before 1972] to 59) have risen from 43% in 1951 to 70% in 1988. During the years of crisis and economic restructuring in the 1970s and 1980s, the change in the gender composition of the labor force was particularly marked. Between 1971 and 1988 male employment as a whole in Britain fell by almost 1.5 million, whereas the number of women in employment increased by more than 1.8 million (Table 5.1).

One of the key reasons for this change has been the growth of jobs in the service sector that has been predominantly a growth of jobs for women, and a decline of manufacturing sector jobs which in the main were filled by men, although women have also lost manufacturing jobs. At the end of the 1980s, 56% of all male workers were service employees, compared with 81% of all female workers and 91% of part-time female workers. The latter factor—the growth of part-time employment for women—has been a particularly marked feature of economic restructuring in Britain (only 7% of men but 42.8% of women workers work part-time). Because of this rise—from 2,757,500 to 4,321,000 part-time jobs for women between 1971 and 1988—the total increase of 1.8 million jobs for women exaggerates the opening of labor market opportunities for women. Rather what the 1970s and 1980s restructuring have achieved is a sharing out of employment between larger numbers of women. Part-time employment seldom brings with it the prospect of economic independence as wages are low (see below) and, in addition, part-time work in Great Britain does not entitle workers to the full range of social benefits that accrue to full-time workers. Eligibility for unemployment and sick pay, for example, is severely limited as are work-related entitlements such as security of employment and holiday pay. In other industrial societies the differences between full-time/part-time entitlements are less marked and, a significantly far lower proportion of women work part-time, although this difference in employment-related benefits may not be the entire explanation. The provision of child care also plays an important part in the structuring of women's employment. For many women in Britain who enter part-time employment, often after working full-time before childbirth, their reentry to the labor market is associated with downward occupational mobility.

The majority of women workers are segregated into a restricted range of occupations in female job ghettoes in the service sector—in catering, cleaning, retailing, secretarial, and clerical work. No less than 42% of all full-time women workers, for example, worked in clerical and related occupations in 1986, while cleaning, catering, hairdressing, and other personal services accounted for 39% of all part-time women workers (Department of Employment New Earnings Survey, 1988). These are the sectors that have expanded most rapidly with economic restructuring, rather than the highly skilled craft jobs that figure so large in the flexible specialization literature.

TABLE 5.2

Ratio of Hourly Earnings of Female to Male Full-Time Employees:
Great Britain, 1971-1989

1971	1972	1973	1974	1975	1976	1977	1978	1979	1980
63.3	63.4	63.2	65.4	69.7	73.2	74.2	72.9	71.3	72.4

1981	1982	1983	1984	1985	1986	1987	1988	1989	
72.6	72.4	75.0	74.2	74.7	74.1	73.4	79.9	76.0	

NOTE: Until 1983 male adult employees were those aged 21 and over, female adults 18 and over; from 1983 the figures are for all those on adult rates. The 1983 estimate on the former basis is 73.1. All figures exclude over-time payments.
SOURCE: Department of Employment, *New Earnings Surveys,* various dates.

Even when women do gain access to relatively well-paid employment on a full-time basis, equality with men remains a chimera. Here too women are segregated into traditionally female semiprofessions such as teaching and nursing (a fifth of all women workers, both full- and part-time, are professional and related workers in education, health, and welfare). Where the rise in women's educational credentials has allowed them to enter "male" professions such as law and accountancy or private sector services such as the business and financial sector, they remain confined to the bottom rungs of the career hierarchy (Crompton & Sanderson, 1990), often held down by the so-called glass ceiling, or restricted to what is known in the United States as "the mommy track." These patterns of gender difference are reflected in the gender segregation of workplaces and in pay differentials. Large numbers of workers work only with individuals of the same sex. For example, more than 50% of all women work in occupations in which three quarters of their coworkers are women. Similarly 50% of all men work in almost exclusively male occupations (more than 90% of their coworkers are men). This gender segregation is even more pronounced at the level of the individual workplace than it is for occupations as a whole (Martin & Roberts, 1984).

Gender differentials in pay rates are also persistent. It has already been suggested that when "traditionally feminine" skills employed by women in the home, such as caring for and servicing others, cooking, cleaning, and sewing, are transferred to the labor market, they are seen as "natural" talents rather than acquired work-related skills and correspondingly are poorly financially rewarded. This tendency is reflected in a comparison of male and female pay rates over time. Despite the rising demand for women's labor as the pool of school leavers shrinks and the implementation of the Equal Pay

Act in 1975—which appears to have effected a once-off, small rise compared with earlier years (although wage restraint policies that held down the rise of male earnings in the early 1970s was also a contributory factor in closing the differential)—full-time female wage rates remain stubbornly at three quarters of the male rate (see Table 5.2).

The increases in women's labor market participation rates have been greatest among women of child-bearing age and particularly marked for women with young children, although there is still a significant difference in participation rates between women with children of different ages. In 1989 41% of mothers of children under 5 were in employment; 12% work full-time and 29% part-time, whereas 69% of women with children between 5 and 9 are in employment (19% work full-time and 50% part-time). The participation rates for women with older children are the same as those for women with no children, that is 79% (figures from the *General Household Survey,* Department of Environment, 1989). Participation rates rose particularly rapidly at the end of the 1980s, after the recession of the earlier years. For example, the rates for women with children under 5 climbed from 24% in 1983, to 36% in 1988, to the 1989 figure of 41%. In addition women as a group are spending longer periods of time in the labor market. For example the average gap between leaving work to have children and returning has declined from seven years at the beginning of the 1970s to just over three years in 1989, and some women, especially those with higher education qualifications, do not leave the labor market at all on child birth, other than for a period of maternity leave. It is thus clear that economic restructuring and women's increased labor market participation rates have implications for total household incomes, the organization of household tasks and for gender relations, although it is difficult to make accurate predictions from aggregate trends. It is hard to untangle the extent to which changes in household and family living arrangements are a cause and/or a consequence of economic changes. What is indisputable is the existence of marked changes in these arrangements.

DEMOGRAPHIC CHANGES

Powerful demographic and social changes in contemporary Britain have challenged notions of traditional family life and, in the last decade, have generated intense political concern about the future of "the family." Briefly stated, women are delaying marriage and childbearing, having fewer children—increasingly outside legal matrimony—and are spending longer periods of their lives living alone or as sole heads of their household, both through

rising longevity and rising divorce rates; all of which are challenging traditional notions of patriarchal authority.

Recent figures from *Population Trends* (OPCS, 1990) allow these
changes to be examined in more detail. Although between 7 and 8 out of
every 10 women eventually marry, the age at which they do so has risen.
Similarly the average age at which women have their first child is rising (25.3
years in 1989, projected to rise to 26.5 by 2000), and the total number of
children born to each woman is falling—it is now less than 2. In addition,
growing numbers of women never have children. Of those born in 1965, now
aged 25, 21% are projected to remain childless, compared with less than 14%
of women in their mothers' generation.

A particularly marked trend, which has fueled intense debate in Britain,
is the rise in the proportion of children born outside marriage. In 1979, when
the Thatcher government first took office, the figure was 10.9% of all births;
by 1989 it had risen to 27%. More than 70% of these births were registered
jointly by the parents, however, and 52% were registered by parents of the
same address—thus indicating that legal definitions and statistical estimates
of "illegitimacy" do not necessarily imply the end of the nuclear family, as
some commentators have assumed. Many of these cohabiting couples who
jointly register births do eventually marry. However, marital breakdown is
also an increasingly common phenomenon. Divorce rates, which were relatively constant in the decades immediately after World War II, have risen
sixfold since 1960 to 12.5% of existing marriages in 1987. At the present
rate, it is estimated that almost 40% of marriages in England and Wales will
end in divorce. According to OPCS (1990) estimates, 7% of all children will
experience the divorce of their parents by their fifth birthday, 12% by the
time they are eight, and 24% by their 16th birthday. Almost all of these
children will live with their mother. The following figures show the family
situation for all children aged 16 and under in 1986.

78% both natural parents, married

2% both natural parents, cohabiting

7% natural mother, stepfather married

2% natural mother, stepfather, cohabiting

10% lone mother

2% other, including single father

SOURCE: *General Household Survey, 1986,* Department of Environment (1989).

Lone parent households now account for 5% of all households. A further 24% of all households are people living alone—many of them women. It is clear that large numbers of women live in households that no longer conform to the idealized notion—that is, a male breadwinner with a dependent wife and children. Only 28% of households are married couples with dependent children, and in a large proportion of these, of course, the woman is employed (figures from *General Household Survey, 1986,* 1989). Both young women and old women are spending longer periods outside the nuclear family. It is an open question whether women's rising labor market participation rates have enabled more women to *choose* to live alone. The combination of occupational segregation and part-time employment outlined above means that for most women, living alone entails a low standard of living or outright poverty rather than economic independence in any real sense. The evidence available from various British poverty studies over the last two decades shows that lone women are overrepresented among the poor. For example in their reanalysis of 1983 Family Expenditure Survey data, Glendenning and Millar (1987) demonstrated that the two types of households most likely to be poor[4] were elderly women living alone and lone mothers (61% of both groups were defined as poor). Together these two groups of women accounted for 15% of all households but 32% of poor households. Further, it is clear that for women living in households with a man in low-income employment or not earning at all, their contribution to total family income is an essential factor in raising these families above the poverty line. For example, it was estimated that, in 1986, the number of families in Britain below the poverty line would have been four times greater than the actual number if women were not in waged employment (Townsend, 1987). Thus with the decline in secure and relatively well-paid employment for men, women's earnings are essential to many dual-headed families in maintaining a standard of living that previously was within reach on the basis of a single wage. It seems that economic restructuring and the consequent increase in low-wage employment in the economy as a whole, for men and for women, may be *increasing* the economic necessity for coupledom rather than facilitating women's independence from men.

The last decade has, however, seen gains for some women. Partly through their increased possession of educational credentials (Crompton & Sanderson, 1986), a small but growing minority of women have gained access to highly paid jobs in the "core" labor market. This access, in combination with the regressive taxation policies of the Thatcher governments, has meant that the 1980s have witnessed growing pay differentials between women. Thus the distribution of earnings between full-time women workers now more

TABLE 5.3

Distribution of Gross Hourly Earnings of

Adult Full-Time Employes, Great Britain

| Ratios to Median | | | | | Women's Wages as a Percentage of Men's Wages | |
| | | Men | | Women | | |
	1980	1989	1980	1989	1980	1989
Top decile	175	200	169	193	71	76
Top quartile	130	140	126	139	72	78
Median	100	100	100	100	73	78
Lowest quartile	81	76	81	78	76	80
Lowest decile	68	60	70	64	76	83

SOURCE: Department of Employment, *New Earnings Survey,* various dates.

closely mirrors that of the distribution between male workers where inequality consistently has been more marked. But as Table 5.3 shows, the distribution for full-time male workers has opened up over the decade, too. The table also reveals that one of the consequences of the entry of women into full-time work accompanied by the loss of manufacturing jobs for men has been that women's pay has not declined relative to that of men in the 1980s. Although gender differentials remain, women, especially those at the bottom end of the distribution, are gaining relative to men in the same decile. The decade was characterized by a new pattern of widening differentials between workers of the *same* sex in addition to the traditional pattern of gender inequalities.

The widening of individual income differentials is reflected in growing disparities in household incomes. Despite the growth in the numbers of women living alone, most women of working age are married or are cohabiting, and, notwithstanding a minority of cross-class marriages, many women marry within their social class. In particular, women in professional occupations tend to marry or live with men in similar jobs. It is these dual-career families that have made the most substantial gains during the 1980s. Among these households, the tasks previously carried out by the now-employed married women are increasingly being purchased: out of the home in the form of fast foods or day care for example, or within the home by the expansion of a range of quasi-domestic service jobs. This commodification of domestic labor has itself increased divisions between women by creating a growing tier of extremely poorly paid service sector employment, almost always for women and often on a casual basis. This type of work, whether in the homes of more affluent women or in the public arena of the "economy," is among the most marginalized and exploitative in the new service economy.

It is only for the small, albeit expanding, minority of women who have gained access to secure, well-paid professional employment and who are relieved from the double burden of routine, boring "women's work" in the home and in the labor market that economic restructuring and the feminization of the labor market is a liberating experience, bringing with it the possibilities for prosperity identified by Piore and Sabel (1984). For these women, the use of their wages to purchase goods and services previously provided at home may have eroded men's power and authority within the home and certainly has expanded women's own choices. The enforced domesticity of the 1950s, identified among middle-class educated women as the "problem that has no name" by Betty Friedan (1963), has disappeared, at least for this group of women. For working-class women, however, unable to purchase even the low paid labor power of other women, the quality of life has deteriorated. Economic necessity has propelled them into the labor market (75% of working-class women compared with 42% of women in the professional and managerial "service" class, according to a recent Mintel (1990) survey, gave "need the money" as their main reason for working) and economic restructuring based on conventional gender stereotypes and continued sex segregation has kept them poor. For most of these women it seems likely that patriarchal power, whether in the home or the labor market, remains relatively untouched. Indeed, even in the more privileged households discussed above, where both partners are in full-time waged employment, in 72% of cases women do most of the domestic work (*Social Trends,* 1989).

It is this combination of women's continued responsibility for domestic labor and for child care, in combination with economic restructuring with the consequent rise in service sector jobs "for women" that has recreated women's marginality, correctly identified but inadequately analyzed by the restructuring theorists criticized above. Gender-specific attributes are created and recreated both in the labor market and in the home. What happens in the latter sphere is not an "adjunct" to labor market restructuring but an essential part of it. While the structural transformation of the economy involves the continual designing and redesigning of jobs and occupations, the new order of the post-Fordist regime is not so much a rupture from previous regimes when women's labor is the central focus, but a continuation and recreation of the same old gender divisions. The labor process and occupational hierarchy may be being restructured but it reproduces and reinforces women's continued location at the bottom end of these structures. To a large extent, the relationship between gender, skill, and pay remains as it was in earlier periods, despite women's growing significance as a group to the economy.

Aggregate analyses, of course, cannot reveal the variation that exists in social conditions between the regions and cities of Great Britain. The particular effects for individual women and their households will depend on the complex interaction of a range of social circumstances. Age, ethnic origin, family life-cycle stage and household composition, housing market position, and geographical location all affect the ways in which the effects of economic change influence the status of women in Great Britain. There are, for example, important differences in the position of women from minority groups, many of whom, especially of Afro-Caribbean origin, are more likely to work full-time than white women as a group, whereas Asian women are more likely to be employed in family enterprises or excluded from the labor market altogether (Bruegel, 1989). There are also differences between black and white women's family circumstances that reflect economic and cultural differences as well as differences because of the younger age profile of minority women as a group. The exact ways in which patriarchal relations in this group intersect with structural socioeconomic changes are complex and require further research. Attitudes to marriage, about sexuality, and early childbirth, for example, vary within and between different ethnic groups (Westwood & Bhachu, 1988). It is clear, however, that the earlier optimistic assumption in feminist theory and in the political practice of the women's movement that entry to the labor market is an essential precursor of women's liberation was incorrect in the majority of cases. For most women entry to the labor market has increased rather than reduced the amount of work that they do, for precious little economic independence. This conclusion is reinforced when changes in the collective provision of welfare services in Great Britain during the 1980s are considered in combination with economic changes. The restructuring of the welfare state is examined briefly below.

THE PATRIARCHAL BASIS
OF SOCIAL WELFARE PROVISION

There is a well-established feminist critique of the patriarchal nature of the welfare state (see, for example, Pateman, 1989) and recent changes in provision have reinforced this. The successive Conservative governments of the 1980s had two aims with respect to welfare provision—first to reduce overall public expenditure in real terms and second, to reshape the system to discourage supposed dependency on state-provided services and benefits. This latter aim was to be achieved not only by increasing reliance on the market but also by increased targeting of benefits on individuals with special needs. The emphasis on the *individual* is particularly important in assessing the ways in which welfare state restructuring has affected the position of women for, as feminist scholars have argued (Balbo, 1987; Land, 1976; Wilson, 1977), the institutions of social welfare provision in Britain assume women's economic dependence on men. Thus the structure of benefits centrally depends on a view that men, through their waged labor or their own entitlement to social income and services, provide for women and children who are dependents.

This notion has continued as a central element in the restructuring of the social security system that has taken place during the 1980s, despite the rhetoric of individual responsibility. Indeed the vision of the patriarchal family as the only socially sanctioned form has been strengthened throughout the period of the Thatcher governments by the unsupported contention from right wing political and social commentators that the availability of welfare benefits and social housing have been a causal factor in the increasing number of mothers living alone. For example, the Centre for Policy studies—the Conservative thinktank—recommended in 1989 that local authorities should stop housing single mothers as it encouraged women to conceive out of wedlock. Further, lone parenthood was blamed as the cause of underachievement at school, juvenile delinquency, crime, and general social disintegration. In October 1990 in the white paper *Children Come First* (HMSO), the establishment of the euphemistically entitled Child Support Agency was announced, in reality as a debt-collection agency. Single mothers will be required to identify the biological father of their child who would then be subject of an earnings attachment order for child maintenance, so reducing single mothers' dependence on the state. Although this change may raise the standard of living of some women, mothers who refuse to identify the father will lose a proportion of their state benefits and so be even worse off than before. In 1988 the Social Security system of Great Britain was restructured when the provisions of the Social Security Act 1986 were introduced. These had the dual purpose of enforcing labor market discipline by decreasing the value of the social wage and reinforcing patriarchal control in the family. The aim was to reduce overall levels of spending by redirecting benefits to those thought to be in "genuine" need. Eligibility for a range of benefits was reduced, as were maximum levels of payment. These changes had a particularly adverse impact on women. For the unemployed, stricter "eligibility for work" rules have been introduced and benefits withdrawn when they are not met. For women with children, their already limited entitlement to this benefit was rendered virtually impossible by a rule that requires the demonstration of already-available childcare facilities without which they were assumed to be unavailable for work. This runs completely counter to established practice, especially in a society in which childcare provision is extremely limited, whether publicly or privately provided (for example, only 28% of the under 5s are in any form of day care in Britain—and this includes "informal" provision such as childminding in the parents' or minders' home—compared with 95% in France). Women with children have also been affected by the decision in 1987 to abolish the annual uprating in child benefit payments, which have remained at their 1987 level since that date.[5] Many pregnant women have lost their entitlement to maternity pay through changes in the scheme, while the universal maternity grant had already been abolished in 1987. Changes in pensions, in both the basic state pension and in the state earnings-related scheme, have affected the living standards of the old, the majority of whom are women. Finally, new benefit regulations that aggregate total *household* income for assessment purposes not only contradict the emphasis on "self reliance" in numerous policy statements and the Green Paper that preceded the 1986 Act, but also assume an equitable division between family members. Studies of family budgets in the United Kingdom have demonstrated that this is seldom the case (Graham, 1984; Morris, 1990; Pahl, 1983).

Women are not only consumers of welfare benefits and services but also essential workers in state provided services and, increasingly, unpaid providers of care "in the community." In Britain such "community care" currently is the orthodox answer to the question of how to care for those who cannot cope alone. While undoubtedly a positive response to criticisms of the restrictions of institutional care, in situations of financial stringency it relies almost entirely on women's unpaid labor, providing care for members of their immediate family in the home (Baldwin, 1985; Finch & Groves, 1983; Green, 1988; Hicks, 1988). In Britain, the White Paper on community care—*Caring for People: Community Care in the Next Decade and Beyond*—which was presented to Parliament in November 1989, established a program of deinstitutionalization in which care for the old, physically and mentally disabled, and terminally ill was to be provided through the market with local authority coordination. Women's voluntary labor was specifically identified as a crucial component in the overall provision. Feminist critiques of community care as a cost-cutting exercise were amply justified when barely seven months after the White Paper's publication, the introduction of the program was postponed until April 1993 in a Treasury public sector budget-cutting exercise. Thus deinstitutionalization, which will continue, combined with no new local service provision will place increasing demands on women at a time when labor market participation rates are increasing. In the context of Britain's aging population, women's responsibility for dependents—first children, then elderly parents, and for the unlucky ones on whom falls the "tricycle of care" (Hardyment, 1990), prematurely dependent husbands—may expand to be a lifetime commitment. Extrapolating from men's partic- ipation in the tasks of domestic labor, it is hard to envisage a significant shift in the gender division of responsibilities for unpaid caring work. Indeed, recent studies (Berthoud, 1988; Glendinning, 1989; Lewis & Meredith, 1988; Millar & Glendinning, 1989) of households in which individuals require considerable care reveal that in almost all cases it is women—both married and those who have never married—who undertake this work. In many cases, these women are forced to leave the labor market and as a result have little financial independence, reliant on income either from the cared-for person, state invalid care benefit (which is paid at an extremely low level), or income transfers from their partners.

RETHINKING FEMINIST THEORY

Before concluding with an assessment of how restructuring has affected women's lives and the relations between men and women, it is appropriate to pause and consider how adequately feminist theories grapple with current changes. One of the most significant achievements of what became known as socialist feminism in the 1970s was its theorization of the connection between housework and waged work, with the insistence that domestic labor was as essential to capitalist accumulation as waged labor. Building on a critique of the gender-blindness of Marxist theory, socialist feminists argued that patriarchal relations, initially identified within the household in the institutions of marriage and the family and in the male control of women's sexuality, are a system or structure of equal importance to capitalism. Each system reinforces the other; indeed patriarchy is seen as essential to the maintenance of capitalist production and its efficient operation. As Zillah Eisenstein argued in her edited collection, *Capitalist Patriarchy and the Case for Socialist Feminism* (1979), capitalism and patriarchy are "an integral process: specific elements of each system are necessitated by the other" (p. 28). Eisenstein based her analysis of the features of patriarchy on the insights of radical feminists who see patriarchy or male supremacy as predating capitalism, although taking a specific form within capitalist societies. She argued that capitalism and patriarchy are neither autonomous systems nor identical, but mutually dependent.

This is not the place to review the long debate that ensued about whether patriarchy and capitalism are dual systems, albeit interdependent, or whether the features of capitalist patriarchy are sufficiently distinct to be a single system of patriarchal capitalism or capitalist patriarchy (see Game & Pringle, 1984; Sargent, 1981; and Walby, 1990 for alternative views). The central element of socialist feminist theory was clear, however—the linchpin was domestic labor that served not only the interests of men but of capital. While men labored in the public sphere to produce surplus value for capital, women labored in the private sphere at reproduction—not only of the men who made things and of children who would be the future workers but also the reproduction of attitudes and capabilities necessary for all kinds of work, including the reproduction of gendered subjects. Thus the home was theorized as an adjunct of the factory, and domestic work was as much a part of the productive process as manufacturing. (The intense and high-level theoretical debate that ensued about whether or not domestic labor produced surplus value generated a large literature but little of lasting value and it is now largely ignored.) The only difference between the man laboring at work and the woman in her home is that the woman is unpaid. A contentious conclusion drawn by some feminists was that wages should be paid for housework, a conclusion disputed by others who argued that this could reinforce women's subordinate position and that the way toward liberation lay in women's full participation in waged labor.

Full participation in the labor market, however, would demand changes in the family—an institution originally seen as in need of abolition, although more recent feminist work has recognized that the family is also a source of support, love, and care in many cases, as well as an arena of male control and sometimes of violence. In the earlier versions of socialist-feminist theory, women's subordinate position in the labor market was seen as a consequence of her domestic position (although the operation of patriarchal relations in the workplace were also quickly identified) that made women a pool or reserve (again the notion of women as a reserve army of labor in the classic marxist sense gave rise to a heated theoretical debate between socialist feminists) of potential workers who could be called in and expelled from the labor force to meet the peaks and troughs of demand.

In light of the social and economic changes of the 1980s a number of these propositions seem in need of reassessment. To take the last point first, it is clear that women are no longer (if they ever were) a pool of disposable labor but are an essential and permanent part of the labor market. But the assumption that is in greater need of reassessment is the notion that capitalism as a system needs domestic labor, or rather is prepared to pay for it through the "family wage." As Ehrenreich (1984) has argued the capitalism-plus-patriarchy model in hindsight assumed a systemic benevolence to capitalism in its willingness to "reproduce labor power" that the events of the last decade have dispelled. As she has written:

> Capital, as well as labor, is internationally mobile, making corporations relatively independent of a working class born and bred in this or any one country. Furthermore capitalists are not required to be industrialist capitalists; they can disinvest in production and reinvest in real estate, financial speculation or, if it suits their fancy, antiques. . . . In their actual practices and policies, capitalists and their representatives display remarkable indifference to the "reproduction of labor power," or, in less commoditised terms, the perpetuation of human life. (p. 273)

And if capitalism does not need domestic labor, what of individual men? As Ehrenreich (1984) wryly remarks, "men have an unexpected ability to survive on fast food and the emotional solace of short-term relationships" (p. 273). As the figures for family breakups and divorces show, "men have been abdicating their traditional roles as husbands, breadwinners and the petty patriarchs of the capitalism-plus-patriarchy paradigm" (p. 273). This is Ehrenreich's interpretation of the situation in the United States. In fact in Britain it is women who are abdicating from their traditional roles as wives as evidenced by the fact that more divorce petitions are now filed by women than by men.

A further indication that the capitalism-plus-patriarchy equation is being rocked by contradiction is the current dilemma of the British state in the face of a welfare system that relies on the old gender divisions that are already receding from view and the restructured economy. Thus white papers and policy documents present a different image of women depending on whether the focus is on social policy or economic productivity and seem unable to resolve the dilemma that this raises for family policy. A theory that enables these contradictions and disjunctions to be comprehended in ways that are sensitive to the specificities of particular places and times is what is required rather than the rather static earlier version of socialist feminism, but retaining the essential insight of the interdependence of the public and private arenas of life.

CONCLUSIONS:
GENDER DIVISIONS/CLASS DIVISIONS
AND THE END OF PATRIARCHY?

It is difficult to assess whether the changes in women's domestic and labor market participation rates have had an impact on the structure of gender relations in Britain (see Allatt, Keil, Bryman, & Bytheway, 1987, for a consideration of the lives of women in different age groups; also Morris, 1990). For women themselves, but perhaps particularly mothers of young children, there obviously have been considerable changes in their lives in the last decades, as they enter the labor market in growing numbers. Waged work brings with it a degree of financial independence, although for many women the net result has been an increase in the total amount of work that they do. On present trends it seems that women's entry into waged labor, and the corresponding increase in their earning power, is an irreversible change. Economic restructuring in contemporary Britain has entailed a general shift toward a low-wage, service-based economy in which the waged work of women as a group is an increasingly central element. Ninety per cent of the increase in the British labor force over the next 10 years is expected to be through women's entry. But for the majority of women it is apparent that restructuring has reinforced, rather than challenged, their segregation into traditional female occupations. Skill shortages (IFF Research Ltd., 1990) and the decrease in the number of 18-year-olds entering the labor market, coupled with women's own aspirations and their growing possession of educational qualifications and career credentials, may effect a once-and-for-all change in the status of women in the labor market. However, as the most recent Equal Opportunities Commission annual report, *Women and Men in Britain, 1990,* concluded, it is "the pitiful lack of childcare which confines most women to part-time, marginalised, low-paid work." Only 2% of children under 5 have full-time places in registered day-care centers and nurseries, whether provided by local authorities or privately. Until high-quality childcare becomes widely available at a price most women can afford, women will remain an exploited group in the labor market.

For many men, too, the consequences of economic restructuring have hardly been positive. The old certainties of secure employment in manufacturing industries have receded and for growing numbers of men marginal or casualized work is becoming more common. In many industrial societies, the United Kingdom and United States among them, capital apparently no longer has any need for the Fordist industrial worker, whose wage labor in a previous era was ensured and reproduced by the domestic labor of an unwaged wife, supported by a "family" wage and the institutions of the welfare state. Increasingly, the marginal groups identified by, among others, Piore, Sabel, Storper, Scott, and Walker—women, rural-urban migrants, migrants from less-industrialized economies and members of ethnic minority groups—are the new model workers of the post-Fordist regime. Their "marginality," in the case of women dependent upon an ideology of femininity and domesticity and a gender division of labor that places the main responsibility for the tasks of reproduction on women, is not just a consequence of economic change but a constitutive element of it. The new slots in the labor market are not devoid of social and gender characteristics but created to draw in particular categories of labor that are easily exploitable. Without an analysis of the domestic position of women, their economic position cannot be fully understood.

Further, economic and social restructuring proceed hand in hand. As Wilson (1988) has rhetorically demanded, "Can it really be coincidence that universal health and welfare services are under attack just as the economy is ceasing to have any use for the old kind of universal workforce?" (p. 198). Coincidence or not, and of course the burden of the argument here is that it is not, it is apparent that the restructuring of the economy and the institutions of the welfare state are based on increasing inputs of women's labor as waged and unwaged workers. This coincidence bears within it a contradiction and possibly the seeds of a crisis. The social "speed up" (Currie, Dunn, & Fogerty, 1980) involved in these increasing inputs is not infinitely extendable. Although individual women so far have borne the brunt of these changes, they are not doing so passively.

Relations in the home are changing. For despite women's "double shift," an inevitable consequence of women's entry into waged labor is that the hours devoted to work in the home are falling (Hartmann, 1987; Jowell, Witherspoon, & Brook, 1986). Patriarchal power relations are surely affected. In addition, a range of struggles—in the workplace and in the community—around so-called women's issues are developing to challenge the old patriarchal certainties and women's oppressed position. Women continue to organize locally, as they have always done, to ensure that some basic level of collective provision, for childcare for example, is available, even if it involves women themselves in various ways caring for each others children. Women are also organizing to improve state-run services and to fight education cuts. In the arena of waged work, women's entry into labor organizations and trades unions is rising, despite the difficulties of organizing part-time workers in many scattered locations. Women's involvement is altering the classic agenda of workplace struggles, often away from issues of pay levels per se and toward terms and conditions of employment. Only a male definition of politics could have ignored these local issues and new forms of organization and bemoaned the decline of working-class action. In other arenas, too, women are organizing to challenge the patriarchal basis of power. For example, there is a growing challenge among Asian women to the tenets of religious fundamentalism and women's exclusion from the public sphere. In London, Women Against Fundamentalism and Southall Black Sisters are active in this struggle. Women's theater, art, and publishing are vibrant and expanding in London and in the other major cities of the United Kingdom.

One of the consequences of economic and social change, however, is that paradoxically both the similarities and the differences between women in contemporary Britain are becoming more marked. As Phillips (1987) has argued,

> The housewife has all but disappeared, the working mother is now the norm, and as far as overall hours spent in paid employment are concerned, there is little to choose between the life of a middle class and working class woman. Most women have children, most run their households without the help of servants. Compared with previous periods the lives of women are now amazingly homogeneous. (p. 62)

But, as argued here, as these similarities have increased, so too have new divisions opened up along class lines. There are growing divides between well-educated women in relatively secure professional employment, with, on paper, pay equity with men and the majority of women in less privileged occupations; between women of different races, between black and white women who in the main do dissimilar jobs in the labor market. These latter divisions in both employment status and in the wider sociocultural sphere during the 1970s and 1980s led to significant splits in the women's movement in Great Britain. Black women have developed a forceful critique of the ethnocentrism of the theories, politics, and practice of the white middle-class women who dominated the movement, especially within the academy and in the control of publishing.

And yet, the reality of gender divisions and of women's oppression still crosses race and class lines. The common experience of childcare is one major uniting factor. The majority of British women have children, albeit fewer and later in life than previously and it is this factor, perhaps above all others, that accounts for women's common experience. And even when women do not have children their gender is seldom an advantage in the labor market. Thus women's experience of waged work, and so their sense of class, remains different from men's. The economic and social restructuring of the 1980s has left untouched this central element in the structure of gender divisions and inequalities between women and men. Despite significant gains in gender equality elsewhere—in women's greater economic self reliance, in educational attainment, in legislation against discrimination and for equal pay for work of equal value—the restructuring of the welfare state, with its central ideological reaffirmation of women's role in the family and their *growing* responsibility for care in the community is a powerful trend countering progress. The whole process of gender construction, the definitions of femininity and masculinity, preconceived notions about the sexual division of labor, both in the home and in the labor market, underpin contemporary changes. Unless gender relations are recognized as a central and constitutive element of the new order, our understanding of these changes will remain partial and incomplete.

NOTES

1. Women themselves often have to be persuaded of the skilled nature of their work and the equity of a comparison with a "male" job before they are prepared to take a case of the industrial tribunal; see Equal Opportunities Commission (1989).

2. In their recent book *The Capitalist Imperative* (1989) Storper and Walker, the proponents of the new industrial geography to be discussed next, include a discussion of the social construction of skills that is remarkable for the complete absence of any mention of gender.

3. Despite the links to the French school of regulationist, there are also differences in these West Coast geographers' work; see, for example, Lovering (1990).

4. Using the commonly accepted definition of poverty as an income of less than 140% of the ordinary rates of Supplementary Benefit—the basic British income support for those ineligible for income-related benefits.

5. In October 1990 it was announced that an additional £1 for the eldest child would be paid from April 1991—an inadequate response to political pressure and the poverty lobby.

REFERENCES

Aglietta, M. (1979). *A theory of capitalist regulation.* London: New Left Books.

Allatt, P., Keil, T., Bryman, A., & Bytheway, B. (Eds.). (1987). *Women and the life cycle.* London: Macmillan.

Amin, A. (1989). Flexible specialization and small firms in Italy: Myths and realities. *Antipode, 21,* 13-34.

Amin, A., & Robins, K. (1990). The re-emergence of regional economies? The mythical geography of flexible accumulation. *Environment and Planning D: Society and Space, 8,* 7-34.

Balbo, L. (1987). Crazy quilts: Rethinking the welfare state debate from a woman's point of view. In A. Showstack Sassoon (Ed.), *Women and the state* (pp. 45-71). London: Hutchinson.

Baldwin, S. (1985). *The costs of caring.* London: Routledge & Kegan Paul.

Baudrillard, J. (1973). *Le miroir de la production.* Tornai, France: Casterman.

Berger, S. & Piore, M. (1980). *Dualism and discontinuity in industrial societies.* Cambridge, UK: Cambridge University Press.

Berthoud, R., (1988) Using benefits to pay for care at home. In S. Baldwin, G. Parker, & R. Walker (Eds.), *Social security and community care.* Avebury, UK: Aldershot.

Bourdieu, P. (1979). *La distinction: Critique sociale du judgement.* Paris: Editions du Minuit.

Breugel, I. (1989) Sex and race in the labour market. *Feminist Review,* No 32, 49-68.

Caring for People: Community Care in the Next Decade and Beyond. (1989). White Paper. London: HMSO.

Central Statistical Office. (1989). *Social Trends, 19.* London: HMSO.

Children Come First. (1990). White Paper. London: HMSO

Cockburn, C. (1983). *Brothers.* London: Pluto.

Cockburn, C. (1985). *Machinery of dominance.* London: Pluto.

Cockburn, C. (1986). Women and technology: Opportunity is not enough. In K. Purcell, S. Wood, A. Waton, & S. Allen (Eds.), *The changing experience of employment* (pp. 173-187). London: Macmillan.

Crompton, R., & Sanderson, K. (1986). Credentials and careers: Some implications of the increase in professional qualifications among women. *Sociology, 20* (1), 25-42.

Crompton, R., & Sanderson, K. (1990). *Gendered jobs and social change.* London: Unwin Hyman.

Currie, E., Dunn, R., & Fogerty, D. (1980). The fading dream: Economic crisis and the new inequality. *Socialist Review, 54,* 102-118.

Department of Employment. (1988). *New Earnings Survey.* London: HMSO.

Department of the Environment. (1989). *General Household Survey Department, 1986.* London: HMSO.

Department of the Environment. (1990). *General Household Survey, 1989,* OPCS Monitor SS90/3. Available from Information Branch, St. Catherine's House, London.

Ehrenreich, B. (1984). Life without father: Reconsidering socialist-feminist theory. *Socialist Review,* No. 73, 48-57.

Eisenstein, Z. (1979). *Capitalist patriarchy and the case for socialist feminism.* New York: Monthly Review Press.

Equal Opportunities Commission. (1989). *Towards equality: A casebook of decisions on sex discrimination and equal pay 1976-1988.* Manchester, UK: Equal Opportunities Commission.

Equal Opportunities Commission. (1990). *Women and men in Britain 1990.* London: HMSO.

Finch, J., & Groves, D. (Eds.). (1983). *A labour of love: Women, work and caring.* London: Routledge & Kegan Paul.

Friedan, B. (1963). *The feminine mystique.* New York: Abacus.

Game, A., & Pringle, R. (1984). *Gender at work.* London: Pluto.

Glendinning, C. (1989). *The financial circumstances of informal carers.* SPRU Report, University of York.

Glendinning, C., & Millar, J. (1987). *Women in poverty in Britain.* Brighton, UK: Wheatsheaf Books.

Gorz, A. (1982) *Farewell to the working class.* London: Pluto.

Graham, H. (1984). *Women, health and the family.* Brighton, UK: Wheatsheaf Books.

Green, H. (1988). *Informal carers: General Household Survey 1985.* London: HMSO.

Hardyment, C. (1990, October 6, 7). Turning over a new age. *The Guardian,* p. 10.

Hartmann, H. (1987). Changes in women's economic and family roles in post-World War II United States. In L. Benaria, & C. Stimpson (Eds.), *Women, households and the economy* (pp. 33-64). London: Rutgers University Press.

Hicks, C. (1988). *Who cares.* London: Virago.

IFF Research Ltd. (1990). *Skill needs in Britain*. London: Author.

Jowell, R., Witherspoon, S., & Brook, L. (Eds.). (1986). *British Social Attitudes 1986 Report*. Aldershot: Gower.

Land, H. (1976). Women: Supporters or supported? In D. Barker & S. Allen (Eds.), *Sexual divisions and society*. London: Tavistock.

Lewis J., & Meredith, B. (1988). *Daughters who care: Daughters caring for mothers at home*. London: Routledge & Kegan Paul.

Lipietz, A. (1987). *Mirages and miracles: The crises of global Fordism*. London: Verso.

Lovering, J. (1990). Fordism's unknown successor: A comment on Scott's theory of flexible accumulation and the re-emergence of regional economies. *International Journal of Urban and Regional Research, 14* (1), 159-174.

Martin, J., & Roberts, C. (1984). *Women and employment: A lifetime perspective*. London: HMSO.

Millar, J., & Glendinning, C. (1989). Gender and poverty. *Journal of Social Policy, 18,* 363-381.

Mintel Survey. (1990). The reasons why women work. Reported in *The Guardian,* September 27, 1990.

Morris, L. (1990). *The workings of the household*. Cambridge, UK: Polity.

Office of Population Censuses and Surveys (1990). *Population Trends*. London: HMSO.

Pahl, J. (1983). The allocation of money and the structuring of inequality in marriage. *Sociological Review, 31,* 237-262.

Pateman, C. (1989). The patriarchal welfare state. In *The disorder of women* (pp. 179-209). Cambridge, UK: Polity.

Phillips, A. (1983). Review of brothers. *Feminist Review, 15,* 101-104.

Phillips, A. (1987). *Divided loyalties*. London: Virago.

Phillips, A., & Taylor, B. (1980). Sex and skill. *Feminist Review, 6,* 79-88.

Piore, M., & Sabel, C. (1984). *The second industrial divide: Possibilities for prosperity*. New York: Basic Books.

Pollert, A. (1988). Dismantling flexibility. *Capital and Class, 34,* 42-75.

Sargent, L. (Ed.). (1981). *Women and revolution: The unhappy marriage of marxism and feminism*. London: Pluto.

Sayer, A. (1990). Post-Fordism in question. *International Journal of Urban and Regional Research, 13*(4), 666-695.

Scott, A. (1988). *New Industrial Spaces*. London: Pion.

Social and Community Planning Research. (1986). *The British social attitudes survey*. London: *Social Trends 19.* (1989). London: HMSO

Storper, M., & Scott, A. (1988). The geographical foundations and social regulation of flexible production complexes. In J. R. Wolch & M. J. Dear (Eds.), *The power of geography: How territory shapes social life* (pp. 21-40). London: Unwin Hyman.

Storper, M., & Walker, R. (1989). *The capitalist imperative*. Oxford: Basil Blackwell.

Townsend, P. (1987). *Poverty and labour in London*. London: Low Pay Unit.

Walby, S. (1989). Flexibility and the changing sexual division of labour. In S. Wood. (Ed.), *The transformation of work?* London: Unwin Hyman.

Walby, S. (1990). *Theorizing patriarchy*. Oxford: Basil Blackwell.

Westwood, S. (1984). *All day every day*. London: Pluto.

Westwood, S., & Bhachu, P. (1988). *Enterprising women*. London: Routledge & Kegan Paul.

Wilson, E. (1977). *Women and the welfare state*. London: Tavistock.

Wilson, E. (1988). *Hallucinations: Life in the post-modern city*. London: Radius.

6

Urban Communities and Crime

RALPH B. TAYLOR

LINKS BETWEEN CRIME and community structure in urban areas constitute the central focus of this chapter. I consider these connections cross-sec-
tionally and longitudinally. Although the bulk of the research has examined
the effects of community structure on crime, I will also note studies investi-
gating impacts of crime on community structure.

The central crime-related issues reviewed are offense rates, victimization
rates, and offending rates of adult criminals or delinquents. *Offense rates*
reflect how many crimes are reported to the police, per 100,000 inhabitants
in a locale. Most of the research on offenses has focused on the "serious" or
Part I offenses that include murder and non-negligent manslaughter, rape,
robbery, aggravated assault, burglary, larceny, motor vehicle theft, and arson.
The FBI routinely publishes counts of recorded crime in *Uniform Crime
Reports.*

Researchers, recognizing that not all victims report crimes to the police,
that police statistics can fluctuate in response to political concerns, and that
crime reporting practices can vary from jurisdiction to jurisdiction, have
backed the development and implementation of a national victimization
survey, the National Crime Survey. Pilot-tested extensively in the 1970s,
implemented in the mid-1970s, and recently redesigned and reimplemented
in the late 1980s, the NCS interviews some 60,000 residents in some
10,000-15,000 households every quarter. Respondents are interviewed, and
reinterviewed five times. From these surveys researchers can gain detailed
information about the incidence of victimizations—the *victimization rate*—
the proportion of respondents victimized during any one time period—the
prevalence of victimization—and explore the circumstances surrounding the
event and its aftermath as well as the characteristics of victims.

Also of interest is information about the perpetrators themselves—the adult offenders and delinquents. *Offending incidence rates* refer to the rate of offenses committed per population count. These can be determined using arrest information from local police departments, or from juvenile court records. *Offending prevalence rates* are constructed by determining the proportion of the population arrested or convicted. *Delinquency prevalence rates* can be constructed by determining the proportion of the youth population adjudicated delinquent. Finally, the effects of criminal justice actions on communities can be explored by looking at the spatial distribution and density of released offenders. These different indices of disorder are summarized in Table 6.1.

This chapter focuses on a number of prevalence and incidence rates (see Table 6.1), as well as dynamics reflective of criminal justice system processing. The different indices examined provide different pictures of the links between crime and community structure. For example, offenders and offenses exhibit markedly different patterns of spatial organization (Baldwin & Bottoms, 1976; Bottoms & Wiles, 1986, 1988) in urban areas. Other dynamics related to crime, such as fear of crime, perception of risk, and behavioral responses such as avoidance, protection, or collective responses are beyond the purview of this chapter.

Recent changes in urban crime rates are examined. These changes reveal some sizable increases over the last 10 years in U.S. cities, but also some decreases. I present four major classes of theories used to examine linkages between community structure and crime, and present typical recent results. The empirical review of recent results relevant to each class of theory is by no means exhaustive. Rather, I attempt to illuminate the contour of findings in each domain.

The four classes of theories considered are: economic, social disorganization/ecological, demographic, and life-style/routine activities. For each class of theory longitudinal as well as cross-sectional information is presented, if available. The subsequent sections briefly discuss recent research viewing crime and criminal justice actions as determinants of community climate and structural properties. The final section summarizes work to date and points toward some possible research futures.

CHANGES IN CITY CRIME

In the nation as a whole, victimization rates for serious, Part I offenses have fallen anywhere from 5% to 30% since the early 1970s (Jamieson &

TABLE 6.1

Central Indices of Crime-Related Issues

Incidence Rates	*Prevalence Rates*	*Other*
Crime rates (reported crimes per unit population)		
Victimization rates: Total number of victimizations reported per unit population, by type of victimization	Victimization prevalence: Proportion of population reporting victimization experiences of a type	
Offending rate: Number of arrests per unit population; can also be self-reported by survey	Offender rate: Proportion of population arrested for offending, or proportion self-reporting offending behavior	
Delinquency rate: Number of delinquent acts recorded per unit of youth population, as indicated by court records or self-reported by survey	Delinquency rate: Proportion of youth population ajudicated delinquent, as indicated by court records	
		Released offender rate: Volume of released offenders per unit population in a community

Flanagan, 1989, Table 3.2). Readers of most big city newspapers would find this surprising. With the advent of crack and the related gang wars in major urban areas, drugs and violent crime seem to be more prevalent than ever before. Particularly senseless violent crimes such as the mugging of the Central Park jogger in New York City in 1989 or the fatal injuries received by a 5-year old boy in a corner store during a "drug-related" shootout in West Philadelphia in 1988, to take just two examples, spark the most shock,

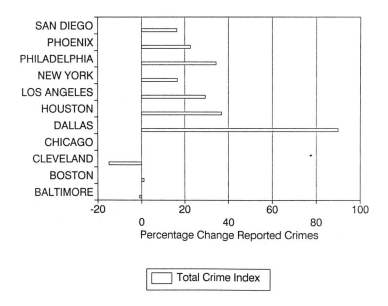

Figure 6.1. Percentage Change Reported Crime Counts (1987-1988) – (1977-1978) 10 Largest Cities

outrage, and concern about the plight of our cities. Can the overall decreasing victimization rates in the country as a whole and the apparently increasing violence in our major cities *both* be occurring at the same time?

Figure 6.1 indicates changes in the total crime index in 9 of the 10 largest cities between 1977-1978 and 1987-1988, for Part I offenses (figures not available for Chicago). The number of crimes reported has declined in some eastern cities such as Cleveland and Baltimore, increased slightly in Boston, and increased substantially in Philadelphia and New York. Increases in southern and western cities range from moderate to sizable. The average increase in the total crime index was 24%.

If we look at reported crime counts for specific crimes such as murder and non-negligent manslaughter, aggravated assault, and burglary (Figure 6.2) we can see that the change in the 10 largest cities varies by crime. Of the three shown, aggravated assault has increased the most (122% on average). Changes in murder averaged a 19% increase, and changes in burglary aver-

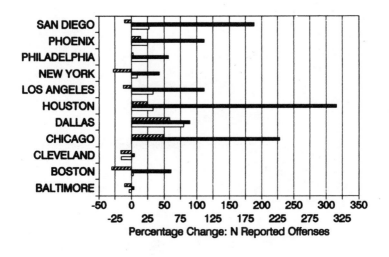

Figure 6.2. Percentage Change in Number of Murders, Aggravated Assaults, Burglaries (1986-1987) – (1977-1978) 10 Largest Cities

aged a 3% increase. The *number* of crimes in our biggest cities are higher in the mid-1980s than they were almost a decade earlier.[1]

If we control for population changes by looking at reported crime *rates,* and expand our scope to look at SMSAs in the United States and other cities, the picture of change is still more of an increase than a decrease, although the percentage changes are not so dramatic. The reported violent crime rate was up 35% in SMSAs and 39% in cities. The reported property crime rate increased less dramatically, up 8% in SMSAs and 16% in cities. Figure 6.3 shows these percentage changes in reported rates for particular crimes between 1977 and 1988.[2] Murder rates were down slightly in SMSAs and cities, and burglary rates were down slightly in SMSAs but up in cities. But robbery, rape, assault, and larceny rates were all up by at least 10%, and in some cases substantially more, in both cities and SMSAs. In short, several serious crime rates were substantially higher in the mid-1980s in urban areas

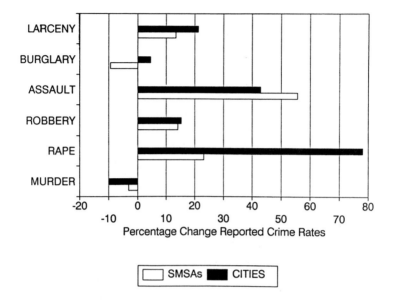

Figure 6.3. Percentage Changes in Urban Reported Crime Rates: 1977-1988

than they were about a decade earlier. Not only has the volume of several serious reported crimes increased; so too has incidence of several of these reported crimes.

THEORIES ABOUT COMMUNITIES AND CRIME, AND EVIDENCE: DETERMINANTS OF DISORDER

This section introduces several classes of theoretical models explaining links between urban structure and crime. Although the bulk of these models have focused primarily on explaining community-level differences in crime rates, some have sought to explain between-city differences. Some of the theories concentrate on explaining static patterns, whereas others seek to explain developments over time.

The theories focus on crime as an outcome. For each class of model I consider its major propositions and key points of empirical support. I review four classes of models: economic, social disorganization/ecological, demographic, and life-style/routine activities.

ECONOMIC MODELS

Stated generally, economic models of crime ask if indices of relative socioeconomic position influence offense and offending rates. Work linking socioeconomic characteristics with offense rates has concentrated on factors such as poverty, relative deprivation, and income inequality. Work linking socioeconomic characteristics with offending rates has focused on unemployment effects and results of programs attempting to reduce unemployment.

Economics and Offense Rates

There are several possible pathways that might link poverty and related structural factors to crime rates. W. B. Miller (1958) combined class and subcultural ideas, arguing that in the lower classes value systems developed encouraging violence as an expression of toughness. Merton (1957) has suggested that poverty encouraged crime because socially acceptable means of achievement were blocked off. Accepting the status-oriented goals of society but denied access to legitimate pathways of achievement, individuals turn to crime. Cloward and Ohlin (1960) extended this anomie theory to cover crimes of violence as well as utilitarian crimes, with their concept of differential access to illegitimate opportunity structures. For an expanded treatment see Rosenfeld (1989).

The latter arguments of Merton and of Cloward and Ohlin suggest that in response to blocked opportunities individuals will experience inner emotional turmoil: frustration, resentment of those better off, or a sense of injustice. These interior states, in response to situations of relative deprivation and combined with available illegitimate opportunities, are postulated as proximate causes of crime.

Along a somewhat different line Blau and Blau (1982; see also Land, McCall, & Cohen, 1990; Sampson, 1985, 1986, 1987) have suggested that situations of relative deprivation can themselves be socially disorganizing. According to this line of reasoning relative deprivation, for example in the form of income inequality, results in crime not only because of the internal psychological states induced in the individuals experiencing the condition, but also because of the deleterious effects on community functioning. Racial and income inequalities widen the gulf between classes and ethnic groups, limit interaction between neighbors, and undermine formal and informal community efforts to control crime. The abrasion of community fabric resulting from psychological and social psychological reactions to relative deprivation plays a major role in increased crime levels according to this variant of the relative deprivation argument.

Research on relative deprivation, the bulk of it conducted using cities or SMSAs as the unit of analysis, has clearly established that indices of lower economic position influence violent crime rates (Blau & Blau, 1982; Land et al., 1990; Rosenfeld, 1986, 1988). Less agreed upon are the relative causal priorities to be attributed to economic factors versus other factors such as southern origin (Messner, 1983), racial composition, racial inequality (Messner & Golden, 1990), or residential mobility, and whether or not relative deprivation or income inequality can "explain away" some of these other effects. For example, Rosenfeld (1986) analyzed SMSA crime rates using a measure of relative deprivation taking into account the intensity of deprivation, scope of deprivation, and level of economic aspirations among poor households. Even after controlling for relative deprivation, which was significantly linked to crime rates, he found that race effects on crime persisted, leading him to conclude that his results provided only "mixed support for the structural model of crime" (p. 127). He also observed regional as well as racial effects, and a stronger link of relative deprivation to crime than of poverty to crime. Rosenfeld (1988) has also been able to link relative deprivation to variations in the national murder rate. With regard to racial inequality the findings have been inconsistent, some suggesting racial inequality, net of relative deprivation, *does* contribute to some homicide rates (Messner & Golden, 1990), others suggesting it does *not* (Land et al., 1990). But the important point, over all, is that at least in developed countries (cf. Rosenfeld & Messner, in press) relative deprivation and homicide are clearly linked.

In addition to questions about the relative importance of economic versus noneconomic factors, the relative importance of relative deprivation, also sometimes called income inequality, versus poverty; and the most appropriate measures of each, have also been debated. Some researchers suggest that poverty is the most pivotal economic factor when considering crimes of murder (e.g., Williams, 1984). For example, Loftin and Parker (1985) used an instrumental variables approach to take account of errors in the poverty variable and linked their poverty measure to homicide rates in the 49 largest cities. Their model worked across different types of homicide. Summarizing the recent literature on poverty and crime they conclude that "the existing literature on crime and economic conditions demonstrates that findings are variable across different levels of aggregation, samples, and model specifications" (p. 281). In the case of poverty they observe: "studies that use multiple indicators of poverty . . . generally find significant positive effects while those that rely on a single proxy variable . . . find weak and inconsistent effects."

Turning to victimization rates, studies find consistent linkages with relative deprivation, but also more complexities than specified by the general theory.

D. A. Smith and Jarjoura (1988) examined victimization rates, using a survey completed in the late 1970s, in artificially defined neighborhoods of three medium-sizes cities. Their findings linked poverty to violent victimization rates, but found that poverty levels interacted with mobility levels in shaping violence levels. "It is the joint occurrence of poverty and a transient population that produces increased criminal violence . . . not all poor areas have high rates of violent crime" (p. 46). Their results point toward a needed theoretical cross-fertilization between structural economic models of violence and social disorganization models.

Findings from an extremely comprehensive study suggest strong links between neighborhood economic level and victimization may be limited to just urban locations. Sampson and Castellano (1982) analyzed the relationship between neighborhood economic status and victimization rates in urban, rural, and suburban areas using *all* data from the 1973 through 1977 National Crime Surveys. Defining neighborhood economic status as the proportion of families in a neighborhood earning less than $5,000 per annum, they found the expected negative relationship between neighborhood economic status and victimization rate operating most strongly in central city locations of SMSAs, weakly in suburban areas, and inconsistently in rural areas. Adult rates of both violent and theft victimization were tied more closely to neighborhood economic status than were juvenile or late teen rates. When neighborhood economic status was defined in terms of unemployment, the effects of economic status on victimization rates were stronger for blacks than for whites. The authors also observed that the contingent link between economic status and crime was observed when age-specific *offending* rates were examined.

Sampson and Castellano (1982) suggested their findings supported a "class crystallization" hypothesis. As one proceeds from rural to suburban to urban areas "residential differentiation in terms of social class and inequality is more salient" (p. 366). In less differentiated rural areas these class differences are not as marked, and thus the predictive power of neighborhood economic status is weakened. This intriguing explanation for the contingent income-victimization linkage does not explain, however, the causal processes whereby the differential ecological salience of class influences victimization and offender rates. Two possible mechanisms would be heightened senses of inequality and injustice among low-income individuals in urban as compared to rural areas, and differential opportunity structures.

In sum, research to date suggests that in urban areas of developed societies such as the United States both poverty and relative deprivation are linked to crimes of violence and instrumental crimes such as thefts. The linkage between these economic variables and crime is clearest in the case of murder and nonnegligent manslaughter. Echoing Loftin and Parker (1985), which factor—relative deprivation or poverty—reveals more explanatory power probably depends in part on the type of crime investigated, the sample used, the level of aggregation, and the specific measures used for the constructs in question. Whether or not these factors vitiate the explanatory power of other cultural (e.g., southern origin), ecological (e.g., mobility), or ethnic factors (e.g., racial composition) probably also depends somewhat on these same factors. Much needed are theoretical models and elaborations of existing models such as Messner and Golden (1990) develop, to help predict and explain (a) contingent linkages that have been observed, and (b) effects varying by homicide type (e.g., white offending versus black offending versus interracial homicide). These interaction effects suggest the effects of economic factors vary depending on other setting conditions such as racial composition and mobility.

Economics and Offender Rates

In its simplest form the question researchers in this area have asked is: What is the relationship between unemployment and crime? Individual-level studies have examined the relationship between participation in job-training programs designed to reduce unemployment. Studies seeking to link such participation with lower subsequent levels of offending have yielded mixed results (McGahey, 1988). Part of the reason for the inconsistent results may be because such training programs do not actually reduce subsequent unemployment levels. The nature of the urban labor market may mitigate against such gains resulting from program participation (McGahey, 1986). But the lack of consistent results stem also from the broad way in which the question is framed (McGahey, 1988). There are many different types of crime, and theoretical models have not yet been developed at this level of specificity.

"Any broad assertion about THE relationship between economic distress and crime is likely to be misleading" (McGahey, 1988, p. 2).

A recent Vera Institute study (McGahey, 1986) explored inner-city youths' perceptions of the economic viability of criminal versus straight careers (p. 250). Youth in minority, low-income neighborhoods perceive the returns from criminal opportunities as substantially higher than the returns from a traditional job. These perceptions, coupled with the paucity of local service and retail employment opportunities in these locations, result in high offending rates among minority youth. Offending rates among white youth may not be as high because they have better networks to connect them to legitimate employment opportunities (p. 247). So the link between local poverty and youth involvement in crime may be moderated by race effects because race is linked with available social ties leading to straight employment opportunities.

In a somewhat parallel vein, studies at the community or metropolitan level of analysis results are pointing toward a relationship between joblessness and crime moderated by race (Sampson & Castellano, 1982). Sampson (1985) looked at race- and age-specific rates of offending across the 53 largest U.S. cities. (The offending rate was derived by multiplying each raw arrest rate by the offense/arrest ratio for the city.) He found that, net of race composition, poverty, and opportunity measures, black offending rates were strongly influenced by level of income inequality, measured with a Gini coefficient. These rates, however, were not influenced by black poverty levels. By contrast, *white* poverty levels had strong positive effects on white violent offending rates. He concludes: "the data support the notion that structural economic factors are important in predicting offending patterns. Specifically, a general pattern emerged in the data whereby income inequality had consistent and relatively strong effects on black offending" (p. 666).

A subsequent cross-sectional study examining race-specific rates of robbery and homicide in 150 large U.S. cities in 1980 begins to uncover the processual dynamics likely to be involved in the economic-offending linkage. Sampson (1987) observed that the effects of black male joblessness on crime were mediated by the effects of joblessness on family disruption. In cities where black unemployment levels were higher, the prevalence of female-headed households in black communities was higher. (He labels a female-headed black household a "disrupted" black household.) In turn, household disruption was linked to black offending rates for murder and robbery; the linkage was especially strong for juvenile offenders. Comparable effects of white family disruption on white violence rates were observed.

These effects persisted after controlling region, income, poverty, race and age composition, city size, and region.

Turning to gang violence and delinquency, Curry and Spergel's (1988) investigation of Chicago's 75 "natural areas" linked poverty levels in Chicago's large communities with gang homicide rate, nongang delinquency rate, and changes in gang homicide rate. They found that the effects of poverty on delinquency and gang homicide were stronger in black as compared to white neighborhoods, and essentially flat in Hispanic neighborhoods.

Sampson's (1987) study is extremely important because it begins to clarify possible pathways linking variations in job market parameters with violence levels. His macro-level study points toward a good number of mediating, micro-level processes at work in these different urban locations. Future work clarifying these macro-to-micro linkages is needed. Findings such as D. A. Smith and Jarjoura's (1988) community-level linkage between percentage single parent households with children 12-20, and victimization rates, and Curry and Spergel's (1988) study of Chicago areas suggest that the city-level dynamics observed by Sampson may be operative and explanatory of differences at the community level.

In sum, cross-sectional studies strongly link joblessness with race- age-specific offending rates at the city level. The effects may be mediated by impacts on family structure and moderated by racial composition. The moderating effects of race may arise from racial differences in connections with social networks leading to employment. Such linkages may also operate at the community level. Studies attempting to predict city-level *changes* in offending rates using joblessness and family structure changes have not yet been completed. One problem impeding the conduct of such studies is the lack of available, accurate city-level demographic information during the years between decennial censuses.

SOCIAL DISORGANIZATION MODELS

Theoretical Processes

Social disorganization models have most frequently been developed within a human ecological framework. In the first half of this century sociologists of the "Chicago School" observed ecological patterning of delinquency rates (Shaw & McKay, 1942, 1972). Rates were higher in the "transition zone"—an area around the expanding central business district where settlement patterns and local social ties were in flux due to the impending encroachment of the central business district. This static geographic pattern has been observed in numerous cities (Harries, 1980). The causal logic posited to bring about higher rates of antisocial behavior was as follows:

(1) Areas experiencing economic deprivation, such as those in the transition zone, or populated by recent immigrants, also experienced high rates of population turnover. As soon as households could, they moved "up and out" of these locations. And, the overall composition of residents in each locale was highly heterogenous.

(2) High turnover rates and heterogeneity made it extremely difficult for residents to organize collectively against groups migrating into the neighborhood.

(3) Because of (1) and (2) the locales were socially disorganized; "the local communities [were unable] to realize the common values of their residents or solve commonly experienced problems" (Bursik, 1988, p. 521).

The mediating processes "carrying" the effects of high turnover and heterogeneity were (a) the difficulty of establishing institutions to foster internal control; (b) difficulties in developing informal local ties (Deutschberger, 1946) and concomitant informal social controls; and (c) blocked channels of communication among groups of residents.

The oft-cited linkage between low socioeconomic status and high delinquency rates (e.g., Gordon, 1967) was interpreted by Shaw and McKay (1972) as a multi-step linkage, involving the above steps (Bursik, 1988, p. 520). Their model can be viewed as "a group-level analog of control theory, and is grounded in very similar processes of internal and external sources of control" (Bursik, 1988, p. 521).

Criticisms of the ecological approach to understanding social disorganization have flourished for quite some time (e.g., Alihan, 1938; Michelson, 1970, chapter 1). Bursik (1988) discusses in detail some of the most cogent criticisms. He points out that researchers have been able to respond to some of these criticisms by modifying their research foci. For example, whereas early researchers assumed an overall urban mosaic that was largely stable, later researchers have examined how neighborhoods or communities can change their role in the larger fabric. Also, whereas early researchers ignored the forces impinging on communities as a result of public policies, recently researchers have begun to incorporate these dimensions. In short, current researchers using an ecological perspective on social disorganization have succeeded in incorporating an expanded range of dynamic and structural factors.

Delinquency Rates

Most of the work in this area has focused on offender rates of juveniles as the outcome of interest. Numerous studies link these rates inversely with areal socioeconomic status, and positively with housing dilapidation (e.g., Shaw, 1929; see also Taylor, 1987 for a review of some of these studies). The most interesting recent studies have examined changes in delinquency rates as a correlate of the unexpected changes, over time, of a neighborhood's position in the larger urban mosaic.

To expand on these latter studies, Bursik (1984, 1986a, 1986b; Bursik & Webb, 1982; Heitgard & Bursik, 1987) has suggested that the portion of a neighborhood's ecological structure at time $T_{(0 + 10 \text{ years})}$ not predictable from its characteristics at T_0, reflects that neighborhood's shifting role in the larger urban ecology during the time frame. The predictable portion of its characteristics at the latter time reflect its earlier fabric and the changes influencing all neighborhoods during the time frame. Stated differently, if a neighborhood is changing more quickly or more slowly over time than other neighborhoods on important ecological characteristics such as racial composition, socioeconomic status, or mobility, its urban role vis-à-vis these other neighborhoods will shift. This role shift will be reflected in $T_{(0 + 10 \text{ years})}$ residuals on the characteristics after regressing scores on the characteristic on T_0 scores.

Bursik's studies of the 75 Chicago natural areas—large communities—during the period 1940 through 1970 found that delinquency rates were linked with these unexpected changes. Unexpected changes in racial composition between 1960 and 1970, reflecting ecological redefinition, were linked with 1970 delinquency rates (Bursik, 1986a; Bursik & Webb, 1982). 1970 delinquency rates were bidirectionally linked with 1970 racial composition in a positive feedback loop (Bursik, 1986b). Further, areas surrounded by neighborhoods undergoing unexpected racial change during the decade also had higher delinquency rates at the end of the period (Heitgard & Bursik, 1987). Perhaps somewhat more surprising was his finding that 1960 delinquency rates were linked with unexpected increases in socioeconomic status (SES) by 1970 (Bursik, 1986b). These Chicago studies of ecological change clearly indicate that a community's shifting role in the urban mosaic influences its delinquency rates.

Two other major studies of changes in delinquency rates, done in Los Angeles and in Racine (WI), also link delinquency rate changes with ecological changes, but in a different fashion than the Chicago studies. Schuerman and Kobrin (1986) examined clusters of census tracts from 1940 to 1970, focusing on those that emerged as high delinquency areas by the end of the period. They found that in the early stages of increasing criminality, physical deterioration preceded rises in delinquency rates, but that at later stages increasing delinquency spurred further increases in deterioration. Parallel to these early stage findings, Shannon (1981) observed that deterioration and out-migration from inner-city areas in Racine was followed by increasing delinquency rates. These non-Chicago studies do not decompose ecological change into predictable and unpredictable portions, thus it is difficult to know how much of these ecological change-delinquency links in Los Angeles and Racine are due to redefinition of community position in the urban mosaic.

In general, these delinquency studies do not address the larger social, political, or cultural forces shaping the pressures operating on urban communities. An exception to this oversight is Bursik's (1989) study of the effects of public housing location on delinquency rates. He observed that public housing in Chicago in the postwar period was located in areas that were unstable to begin with, the placement subsequently resulted in even higher instability levels in these communities, and delinquency rates were heightened as a consequence of the higher instability levels.

A second shortcoming of these studies is that they do not directly address the micro-level dynamics of informal social control posited to emerge from these ecological changes and result in elevated offender rates. One exception is Simcha-Fagan and Schwartz's (1986) study of 500+ adolescents in 12 New York City neighborhoods. In addition to collecting census data they interviewed mothers about informal and formal social controls in the neighborhoods. They found that contextual characteristics were significantly related to delinquency, but explained only 2% - 4% of the outcome variance. They concluded that the effects of neighborhood context on delinquency were probably indirect, and mediated by the socialization experiences occurring in the community. Since they had only 12 neighborhoods in their study these null or small effects should not be considered damaging to the social disorganization viewpoint. The effects they observed may be regarded as a lower bound for such effect sizes.

In sum, studies link ecological shifts with changes in delinquency rates. The roles and impacts of hypothesized changes in local social networks, informal social controls, or opportunity structures resulting from these ecological shifts have not yet been established. Work on the precursors of these forces leading to change, and how these forces are channeled into particular locations, is just beginning.

Crime Rate Changes

A pair of studies examining Baltimore (MD) neighborhoods in the 1970s have linked degree of unexpected ecological change with degree of unexpected reported crime change. These studies suggest that neighborhoods undergoing *rapid* ecological change will experience increasing disorder, even if that change is generally regarded as a "positive" development, reflecting neighborhood improvement.

Taylor and Covington (1988, 1990) focused on neighborhoods that, at the beginning of the decade, were likely, due to their status, stability, and race and youth composition, to experience gentrification or further entrenchment into an underclass status during the period. Unexpected changes in ecological structures of these neighborhoods during the period, reflecting their shifting roles in the larger urban mosaic, were linked with unexpected changes in violent crime rates.

In examining neighborhoods being further absorbed into underclass status the authors suggested the following: if unexpected economic changes were most closely linked to violence changes, the pattern would support a relative deprivation explanation of the linkage, whereas if unexpected changes in stability were most closely linked to changes in violence, a social disorganization interpretation of the linkage would be most warranted. The pattern of results for murder was most supportive of a relative deprivation explanation; the economic dimension of ecological change was most closely linked to the changes in murder. The data hinted that in some neighborhoods moderate levels of stability may have "buffered" the emergence of increasing murder levels as economic status declined.

By contrast, when changes in aggravated assault rates were considered, the observed pattern indicated that changes in stability were more closely linked to changes in violence, than were changes in economic status. Thus the pattern for assault changes more closely supported the social disorganization perspective. Turning to gentrifying neighborhoods, unexpected increases in *stability* were more closely linked to unexpected changes in assault and murder rates—in the positive direction—than were changes in economic level. Although the pattern of results would seem to support a social disorganization perspective more strongly than a relative deprivation viewpoint, many of the gentrifying neighborhoods were in close proximity to public housing sites or extremely low-income neighborhoods, both of which may have provided sizable pools of offenders whose sense of relative injustice was inflamed by nearby developments.

In another study examining gentrifying neighborhoods more closely and looking at "rational" crimes of larceny and robbery, Covington and Taylor (1989) found that these neighborhoods may have contained their own sources of potential offenders. The fabric of neighborhoods in Baltimore that had been gentrifying for a time was quite mixed. Persistent pockets of low-income households and poverty remained. Parallel textures have been observed in other gentrifying neighborhoods in other cities such as Nashville and Philadelphia. Before gentrification many of these neighborhoods had been locations where criminal subcultures could readily develop. After prolonged gentrification these neighborhoods still retained segments of this earlier composition. Gentrification is a spatially "spotty" process. In more traditional ecological terms, the invasion-succession cycle, in contrast to expectations, can "stall." Important questions include: What factors—locally, regionally, and nationally—lead to such a stall? Is it possible to build community in these highly diverse areas? Are local, informal social controls likely to develop?

Consistent with the persistently diverse composition of gentrifying neighborhoods, Covington and Taylor observed that degree of gentrification—as measured by unexpected increases in relative house value—was linked with unexpected increases in larceny and robbery rates. Gentrifying neighborhoods, as compared to neighborhoods appreciating less dramatically, experienced more sizable positive unexpected changes in larceny and robbery rates. Burglary, however, declined somewhat more strongly in gentrifying as compared to less dramatically appreciated neighborhoods. The pattern of findings in gentrifying neighborhoods for rational crime suggests that rational offender processes—explained below in the section on routine activity theories—may mediate the connections between unexpected ecological change and unexpected crime changes.

In toto, these Baltimore studies suggest the following. First, rapid ecological change, regardless of the direction of the change, is associated with increasing disorder. Neighborhoods whose role in the urban mosaic is more radically redefined during the period, vis-à-vis other neighborhoods, were more at risk of increasing crime. The linkage of increasing disorder with swift ecological change is broadly supportive of the ecological perspective on social disorganization, although the relevant mediating pathways have not been tested as yet. By implication: If local power structures, private and public, have the ability to prevent rapid ecological change in or near urban neighborhoods at risk of increasing disorder, the exercise of this ability may result in lower crime rates in these locations. Second, it is possible, within an ecological framework, to dovetail an ecological perspective with other theoretical models and derive sensible predictions. In explaining violence changes we use relative deprivation theory. In explaining rational crimes we used a rational offender perspective (Clarke & Cornish, 1986). Consequently, we would suggest that it is possible to formulate broader models of spatial structure and disorder that incorporate compatible elements from the ecological perspective and other viewpoints.

Victimization and Offender Rates

A major study of Sampson and Grove (1989) begins to link ecological features of neighborhood fabric with features of local social climate, and the latter to victimization rates. Using the 1982 and 1984 British Crime surveys they aggregated data to the large areas used as a sampling level. They argued that community SES and ethnic heterogeneity were linked to victimization rates because they resulted in unsupervised local teen corner groups, and a paucity of local friends. The community-level analyses of personal victimization and offender rates supported their contention. Such results provide strong cross-sectional support for the ecological model of social disorganization. The next steps required to further support this model would be to examine such connections *longitudinally,* and to use small, ecologically valid neighborhoods rather than larger, politically driven community boundaries.

Summary

The ecological model of social disorganization has received cross-sectional support (e.g., Sampson & Grove, 1989; Simcha-Fagan & Schwartz, 1986; but cf. Messner & Golden, 1990). The model appears broadly applicable to offender rates, usually of juveniles; reported crime rates; and victimization rates. Longitudinal studies have connected key exogenous parameters and outcomes, but have not yet substantiated the mediating pathways involving community social control and climate. The model has been successfully dovetailed with other theoretical orientations. Work explicating the social, cultural, and political origins and targeting of the ecological forces acting on neighborhoods has not yet begun in earnest.

DEMOGRAPHIC MODELS

Of late the bulk of researchers have viewed demographic variables *within* a routine activities or life-style model of disorder. But there are some researchers whose links between demographic characteristics and disorder do not fit within such a model. I briefly review a recent study of this ilk here.

The age-crime linkage has received considerable attention in recent years. Researchers have observed that peak offending years are the teen years and early twenties and peak incarceration years are the twenties and early thirties. Chilton (1986) offered a particularly intriguing recent application of this link in his explanation of urban crime rates. He compared 1960 and 1980 arrest rates, by age, sex, and race groups, and argued that sociodemographic changes in age composition in large cities could be used to explain crime rate changes. Using changes in demographic composition from 1960 to 1980, he was able to predict 53% of the increase in arrest rates for different age-sex-race categories in the 12 largest U.S. cities. Increases in the young, male, nonwhite populations in these cities explained sizable amounts of these predicted arrest changes. He recognized, however, that this demographic-crime linkage was contingent upon the larger economic changes occurring during this period (p. 115). In short, underpinning the age change-crime change link were worsening economic prospects for particular segments of the population in these large cities.

ROUTINE ACTIVITIES OR LIFE-STYLE THEORIES OF DISORDER

Life-style theories of victimization, originally promulgated by Hindelang, Gottfredson, and Garofalo (1978), explained victimization rates as resulting from the victim's proximity, in space and time, to potential offenders. Later recast in a more structural vein by Felson, Cohen, and others, and dubbed a "routine activities" theory of crime, the latter version has three central propositions. For a crime to occur there are three prerequisites: the presence of a motivated offender, an absence of a capable guardian, and a suitable target.

This theoretical kernel has led to a variety of findings. For example, Cohen and Felson (1979) hypothesized that increasing dispersion away from the household, such as has occurred in the post-World War II era, has resulted in higher opportunities of crime and thus higher crime rates.

Several studies have applied the model to victimization rates (e.g., Cohen & Cantor, 1981; Felson, 1986; Garofalo, 1987; Land & Felson, 1976; Maxfield, 1987; and Sampson & Lauritsen, 1990). For example, Cohen, Kluegel, and Land (1981) used five risk factors to explain criminal victimization levels in the 1974 and 1977 NCS: exposure, guardianship, proximity to potential offenders, attractiveness of potential targets, and "definitional properties of the crimes themselves" (p. 505). Their measure of proximity to potential offenders was central city location *and* living in a low-income neighborhood. Proximity to potential offenders was linked to higher victimization risk for burglary, assault, and larceny.

The routine activities approach leads to a reinterpretation of social class-victimization and demographic-victimization links observed in other studies using structural models. For example, Cohen et al. (1981) argued that "those usually thought to be the most vulnerable economically and socially—the poor, nonwhite, the old—are not the most likely victims of crime. Race has little direct effect on victimization risk" (p. 505). According to this model, demographic or social class characteristics lead to certain activity routines and locations that, in turn, influence victimization rates.

Miethe, Stafford, and Long's (1987) analysis of the 1975 NCS of 13 major cities provides an excellent example of the reinterpretation of structure-victimization links afforded by the routine activities perspective. They found that interactions of sociodemographic characteristics with activity variables helped explain property crime victimizations but not violent crime victimizations. They also found that to some extent life-style factors mediated sociodemographic effects because some of the latter were rendered nonsignificant after adding the former.

Another example is provided by Maxfield's (1987) analysis of the British Crime survey and the 1983 Victim Risk Supplement to the NCS. He found that single parent households were at risk of violent crime because (a) they were most likely to be victimized by present and former spouses, and (b) often being unemployed, they spent much time in the home domain where they were proximate to their present or former spouse. Maxfield's findings point up the importance of focusing on specific domains and specific crime-victim-perpetrator clusters when interpreting demographic and life-style connections with victimization.

From the viewpoint of structure and disorder, four shortcomings of routine activities models merit attention. First, measures of proximity to potential offenders have been extremely gross and quite indirect. Offender density and structural characteristics of areas with high offender rates have been confused. Second, as Miethe et al. (1987) have pointed out, tests to date have explained individual-level outcomes. It is not at all clear to what extent the model explains between-community variations in victimization rates. Third, the model assumes a rational offender perspective (Clarke & Cornish, 1986). This may be decreasingly applicable over time (Miethe et al., 1987), or may have varying applicability depending upon the type of crime or the crime-victim-perpetrator constellation. If and as structural conditions worsen in urban areas offenders may become increasingly motivated by nonrational factors such as reactions to relative deprivation. Assumptions are made about

⸜otivations whose accuracy is not known. And finally, the societal and processual dynamics creating connections between demographic characteristics and routine activity patterns remain unexamined.

EFFECTS OF CRIME ON STRUCTURE

What are the impacts of crime rates on aspects of community structure such as stability, house value, and so on? I summarize here some of the major studies done on these topics.

MOBILITY

It seems plausible to expect that residents living in higher crime neighborhoods will be more desirous of leaving, and in fact will be more likely to migrate out of the high crime neighborhood. A major study done in the 1960s confirms the former connection, but not the latter. Using perceptions of crime and violence rather than actual crime or victimization rates, and interviewing 1,400+ households in 43 metropolitan areas, Droettboom, McAllister, Kaiser, and Butler (1971) found that perceptions of disorder were more strongly linked to desire to move than they were to actual mobility. Perceptions of serious local violence were not connected with extra-neighborhood moves, although those perceiving more crime in their locale were more mobile. The authors suggested that many constraining factors may have prevented those who perceived crime as a serious problem in their locale from being able to migrate out. Further, this connection between intent to move and perceived local crime was considerably stronger in central city as compared to suburban locations.

Droettboom et al. (1971) were also unable to confirm the hypothesis that high crime leads to a suburban exodus. The proportion of movers leaving central city locations and ending up in suburban locations 3 years later who saw crime in their old neighborhood as a serious problem was not different from those who viewed crime in their neighborhood less seriously. "It does not appear that the perception of crime and violence results in a migration from central city to suburb" (p. 323). But again, it is hard to know the role played by economic constraints in shaping this relationship. Those living in higher crime neighborhoods are less likely to have the means to move, as compared to those living in lower crime neighborhoods.

The lack of crime's impact on urban out-migration observed by Droettboom et al. with individuals was later confirmed by Frey's (1979) aggregate analysis of 39 large SMSAs in 1970. He conceptualized the city-to-suburb migration stream as a two-stage process: "(1) the decision to move, and (2) the choice of destination" (p. 429). He observed no effect of crime rate on mobility incidence (rates), but did observe some effects of crime on the selection by recent movers of suburban destinations. A very high correlation in the sample (.41) between proportion of central city population which was black, and central city crime rates, undoubtedly limited the size of the crime effects he could observe.

More recently, Taub, Taylor, and Dunham (1984) interviewed owners and renters in eight neighborhoods in Chicago. They found homeowners, living in very high crime neighborhoods, insisting they would stay in the locale. These owners were planning on staying in large part because their home investment was appreciating at the time. This study was conducted in the early 1980s when house value appreciation was occurring in revitalizing neighborhoods in Chicago and other cities. Whether such persistence would be evident in the early 1990s when house values are not appreciating in these locations has not been determined.

Although all of these studies to date are somewhat limited, they fail to support the notion that high crime levels in urban areas result in higher suburban-ward mobility rates. High crime areas are, however, often high mobility areas, as the social disorganization theorists have observed (Kornhauser, 1978). Analyses delineating how at the neighborhood level crime elevates mobility, and vice versa, have not yet been completed.

HOUSE VALUES

More clear-cut are links between local crime rates and house values. E. S. Miller (1981) has outlined the processual dynamics involving real estate agents, residents, buyers, and local lending institutions that contribute to this linkage.

The work to date shows clear impacts of local disorder rates on house values. Work in Boston (Hellman & Naroff, 1979; Naroff, Hellman, & Skinner, 1980; see also Gray & Joelson, 1979) finds sizable effects of crime on house values, suggesting that a 5% decrease in crime could result in 7-30 million dollars in increased tax revenue. Dubin and Goodman (1982) examined effects of local crime rates on Baltimore City and Baltimore County house sale values, controlling for structural housing features. Sizable effects of crime on house price were observed in both locations, with larger effects evident in the city. For example, in the city a one standard unit increase on the violent crime component implied a drop of more than $3,000 in house value. This study also controlled for educational opportunities in each location. In sum, the work to date in this area clearly points toward sizable effects of crime on local economic structure in the form of house values.

EFFECTS OF CRIMINAL JUSTICE ACTIVITIES ON NEIGHBORHOOD STRUCTURE

In the 1970s and 1980s the criminal justice system has been releasing large numbers of offenders and ex-offenders into urban communities. Some of those released are serving their sentence in the community and are under probation or parole supervision; others are released after having served a sentence of incarceration. A couple of recent studies have documented effects of these criminal justice activities on neighborhood fabric.

Gottfredson and Taylor (1988) followed up a cohort of several hundred ex-offenders released to a sample of Baltimore City neighborhoods in the early 1980s. Controlling for sociodemographic characteristics of the neighborhood, they observed impacts of released ex-offender density on residents' fear of crime, attachment to the neighborhood, and confidence in the future of the neighborhood. Although released offenders clearly have a right to return to the communities in which they resided before they were incarcerated, these release patterns are having substantial effects on residents' perceptions of the safety and viability of their home neighborhood.

Rengert (1989) examined the spatial distribution of recidivist burglar offender rates and burglary offenses in Philadelphia. The offender rates were considerably higher in central city locations, as were the offense rates for burglaries. He also observed that more burglars were traveling *into* these central city locations to commit their offenses, than were traveling into outer city locations. Rengert concludes that "those in our population least able to bear the high physical, economic and emotional burden of property crime are victimized most by recidivist burglars . . . the poor in our society are victimized disproportionately by the criminal justice system when undersupervised convicts are released" (pp. 557-558).

Criminal justice policies, such as the releasing of nonviolent offenders in Philadelphia in order to alleviate prison overcrowding, have spatial and thus socioeconomic ramifications. Supervision policies for released offenders or efforts to better integrate the returning offender into the community might help reduce the disproportionality of some of the burdens associated with high released offender density. It is not clear at this time, however, what policies would be effective in achieving this goal. Even policies thought to provide a high degree of community safety, such as electronic monitoring, may allow up to half of the participants to escape supervision for long periods of time (Baumer, Mendelson, & Maxfield, 1990). Initiatives such as electronic monitoring may allow releases in the future to increase substantially (Berry, 1985), turning communities into electronic jails (Harland & Rosen, 1987)

Criminal justice policies play roles in shaping rates of returned offenders in communities. Future actions of these returned offenders can also be influenced by policymakers through delivery of various services (Savelsberg, 1984). Returned offender rates have sizable impacts on aspects of community life related to disorder, confidence, and residential attachment and satisfaction.

CONCLUSIONS AND RESEARCH FUTURES

Levels of disorder—offender rates, offense rates, and victimization rates—are higher in central city locations than they are in suburban locations. These levels have been increasing in the largest cities over the last decade despite nationwide drops in victimization rates. Levels of disorder are higher in some types of urban communities as compared to others. Communities that we typically think of as disadvantaged—having more black or Hispanic households, and/or more lower income households, and/or more unstable or disrupted households—have higher disorder levels. Structural explanations of this connection need to consider the higher levels of offenders and potential victims (opportunities) in these locations (S. Smith, 1986) as well as the roles of criminal justice policies. The processual dynamics linking these sociodemographic community characteristics to higher crime and victimization rates are interpreted differently by different theoretical models. Relative deprivation models focus on psychologically induced states or intergroup attitudes; social disorganization models focus on group-level control dynamics; differential association or opportunities theories focus on social and other contexts with which the potential offender is confronted; and routine activities models focus on space-time behavior patterns of victims in relation to potential offenders. Results to date do not permit electing one model over another as a superior theoretical kernel for explaining the links between disorder and community structure.

In general terms work to date exhibits two major shortcomings. First, studies have either examined aggregate (SMSA or community) or individual dynamics. Few studies have spanned both levels and allowed us to unpack the connections crossing these levels. Baldwin (1979) voiced this same lament more than a decade ago. Researchers need to develop multi-level models incorporating micro-level dynamics in response to community-level structures and changes. These models should probably be somewhat domain specific (e.g., the outdoor residential environment, small commercial centers, etc.) For example, Taylor (1989) has made an attempt to develop such a model for streetblocks in the urban residential environment, linking neighborhood dynamics with streetblock functioning and individual-level outcomes.

Second, studies conducted on communities within an urban area have failed to explore the cultural and political forces acting on communities and the roles of these forces as shapers and channelers of ecological change. Although there are some exciting exceptions (e.g., Bursik's [1989] study showing effect of public housing location on subsequent neighborhood instability), the forces behind community impacts and changes have been neglected. Part of this neglect is due to early ecologists' conceptualization of these larger forces as "subsocial" (Michelson, 1970, chapter 1). But the ecological orientation does not mandate such a viewpoint, nor does it exclude the origins of these forces from examination.

One study probably cannot respond to both of the above shortcomings. Rather, I envision future work in this area as pursuing either one focus (extra-community origins of forces influencing community fabric and disorder) or the other (linking community structure and changes to mediating structures and individual-level outcomes).

Future work will also, we hope, continue to expand upon some of the most exciting features of work to date including an examination of longitudinal dynamics as communities change, and a consideration of the contributions of disorder and criminal justice policies to community life and populations, as well as examining the effects of community on disorder. There are only a few extant theoretical models to guide such work (see Taylor & Gottfredson, 1986, for a review). Modeling efforts incorporating insights from the different theoretical perspectives reviewed here are desperately needed.

NOTES

1. These figures were calculated from Flanagan, Hindelang, and Gottfredson (1980), Table 3.119; Flanagan and Maguire (1990), Table 3.122.

2. These figures were calculated from Flanagan, Hindelang, and Gottfredson (1980), Table 3.65; and Flanagan and Maguire (1990), Table 3.119.

REFERENCES

Alihan, M. A. (1938). *Social ecology: A critical analysis*. New York: Columbia University Press.
Baldwin, J. (1979). Ecological and areal studies in Great Britain and the United States. In N. Morris & M. Tonry (Eds.), *Crime and justice: An annual review of research* (Vol. 1). Chicago: University of Chicago Press.
Baldwin, J., & Bottoms, A. E. (1976). *The urban criminal*. London: Tavistock.
Baumer, T., Mendelson, R., & Maxfield, M. (1990, November). *Effects of home confinement.* Paper presented at the annual meetings of the American Society of Criminology, Baltimore.
Berry, B. (1985). Electronic jails. *Justice Quarterly, 2,* 1-22.
Blau, J., & Blau, P. (1982). The cost of inequality: Metropolitan structure and violent crime. *American Sociological Review, 47,* 114-119.
Bottoms, A. E., & Wiles, P. (1986). Housing tenure and residential community crime careers in Britain. In A. J. Reiss, Jr., & M. Tonry (Eds.), *Communities and crime* (pp. 101-162). Chicago: University of Chicago Press.
Bottoms, A. E., & Wiles, P. (1988, November). *Urban areas, social change and crime.* Paper presented at the annual meeting of the American Society of Criminology, Chicago.
Bursik, R. J., Jr. (1984). Urban dynamics and ecological studies of delinquency. *Social Forces, 63,* 393-413.
Bursik, R. J., Jr. (1986a). Ecological stability and the dynamics of delinquency. In A. J. Reiss & M. Tonry (Eds.), *Communities and crime* (pp. 35-66). Chicago: University of Chicago Press.
Bursik, R. J., Jr. (1986b). Delinquency rates as sources of ecological change. In J. M. Byrne & R. J. Sampson (Eds.), *The social ecology of crime* (pp. 63-76). New York: Springer.
Bursik, R. J., Jr. (1988). Social disorganization and theories of crime and delinquency. *Criminology, 26,* 519-551.
Bursik, R. J., Jr. (1989). Political decision making and ecological models of delinquency: Conflict and consensus. In S. F. Messner, M. D. Krohn, & A. E. Liska (Eds.), *Theoretical integration in the study of deviance and crime: Problems and prospects* (pp. 105-117). Albany: SUNY Press.
Bursik, R. J., Jr., & Webb, J. (1982). Community change and patterns of delinquency. *American Journal of Sociology, 88,* 24-42.
Chilton, R. (1986). Age, sex, race and arrest trends for 12 of the nation's largest central cities. In J. M. Byrne & R. J. Sampson (Eds.), *The social ecology of crime* (pp. 102-115). New York: Springer.
Clarke, R. V., & Cornish, D. B. (1986). Modeling offenders' decisions: A framework for research and policy. In N. Morris & M. Tonry (Eds.), *Crime and justice: An annual review.* Chicago: University of Chicago Press.
Cloward, R. A., & Ohlin, L. E. (1960). *Delinquency and opportunity*. New York: Free Press.
Cohen, L., & Cantor, D. (1981). Residential burglary in the United States: Lifestyle and demographic factors associated with the probability of victimization. *Journal of Research in Crime and Delinquency, 18,* 113-127.
Cohen, L. E., & Felson, M. (1979). Social change and crime rate trends: A routine activity approach. *American Sociological Review, 44,* 588-608.

Cohen, L. E., Kleugel, J. R., & Land, K. C. (1981). Social inequality and predatory criminal victimization: An exposition and test of a formal theory. *American Sociological Review, 46,* 505-524.

Covington, J. C., & Taylor, R. B. (1989). Gentrification and crime: Robbery and larceny changes in appreciating Baltimore neighborhoods in the 1970's. *Urban Affairs Quarterly, 25,* 142-172.

Curry, G. D., & Spergel, I. A. (1988). Gang homicide, delinquency and community. *Criminology, 26,* 381-405.

Deutschberger, P. (1946). Interaction patterns in changing neighborhoods. *Sociometry, 9,* 303-315.

Droettboom, T., McAllister, R. J., Kaiser, E. J., & Butler, E. W. (1971). Urban violence and residential mobility. *Journal of the American Institute of Planners, 37,* 319-325.

Dubin, R. A., & Goodman, A. C. (1982). Valuation of education and crime neighborhood characteristics through hedonic housing prices. *Population and Environment, 5,* 166-181.

Felson, M. (1986). Linking criminal choices, routine activities, informal control, and criminal outcomes. In D. Cornish & R. V. Clarke (Eds.), *The reasoning criminal: Rational choice perspectives on offending* (pp. 119-128). New York: Springer.

Flanagan, T. J., Hindelang, M. J., & Gottfredson, M. R. (1980). *Sourcebook of criminal justice statistics: 1979.* Washington, DC: Government Printing Office.

Flanagan, T. J., & Maguire, K. (Eds.). (1990). *Sourcebook of criminal justice statistics: 1989.* Washington, DC: Government Printing Office.

Frey, W. H. (1979). Central city white flight: Racial and non-racial causes. *American Sociological Review, 44,* 425-448.

Garofalo, J. (1987). Reassessing the lifestyle of criminal victimization. In M. Gottfredson & T. Hirschi (Eds.), *Positive criminology* (pp. 23-42). Newbury Park, CA: Sage.

Gordon, R. A. (1967). Issues in the ecological study of delinquency. *American Sociological Review, 32,* 927-944.

Gottfredson, S. D., & Taylor, R. B. (1988). Community contexts and criminal offenders. In T. Hope & M. Shaw (Eds.), *Crime and community context* (pp. 62-82). London: HMSO.

Gray, C. M., & Joelson, M. R. (1979). Neighborhood crime and the demand for central city housing. In C. M. Gray (Ed.), *The costs of crime* (pp. 47-60). Beverly Hills, CA: Sage.

Harland, A., & Rosen, C. J. (1987). Sentencing theory and intensive supervision probation. *Federal Probation, 51,* 33-42.

Harries, K. D. (1980). *Crime and the environment.* Springfield, IL: Charles C Thomas.

Heitgard, J., & Bursik, R. (1987). Extracommunity dynamics and the ecology of delinquency. *American Journal of Sociology, 92,* 775-787.

Hellman, D. A., & Naroff, J. L. (1979). The impact of crime on urban residential property values. *Urban Studies, 16,* 105-112.

Hindelang, M. J., Gottfredson, M. R., & Garofalo, J. (1978). *Victims of personal crime: An empirical foundation for a theory of personal victimization.* Cambridge, MA: Ballinger.

Jamieson, K. M., & Flanagan, T. J. (1989). *Sourcebook of criminal justice statistics: 1988.* Washington, DC: Government Printing Office.

Kornhauser, R. R. (1978). *Social sources of delinquency.* Chicago: University of Chicago Press.

Land, K. C., & Felson, M. (1976). A general framework for building dynamic macro social indicator models: Including an analysis of changes in crime rates and police expenditures. *American Journal of Sociology, 82,* 565-604.

Land, K. C., McCall, P. L., & Cohen, L. E. (1990). Structural covariates of homicide rates. *American Journal of Sociology, 95,* 922-963.

Loftin, C., & Parker, R. N. (1985). An errors-in-variables model of the effect of poverty on urban homicide rates. *Criminology, 23,* 269-285.

Maxfield, M. G. (1987). Lifestyles and routine activity theories of crime: Empirical studies of victimization, delinquency and offender decision-making. *Journal of Quantitative Criminology, 3,* 275-282.

McGahey, R. M. (1986). Economic conditions, neighborhood organization, and urban crime. In A. J. Reiss & M. Tonry (Eds.), *Communities and crime* (pp. 231-270). Chicago: University of Chicago Press.

McGahey, R. (1988). *Jobs and crime* (National Institute of Justice CRIME FILE Study Guide). Washington, DC: National Institute of Justice.

Merton, R. K. (1957). *Social theory and social structure.* Glencoe, IL: Free Press.

Messner, S. F. (1983). Regional and racial effects on the urban homicide rate: The subculture of violence revisited. *American Journal of Sociology, 88,* 997-1007.

Messner, S. F., & Golden, R. M. (1990, November). *Racial inequality and racially disaggregated homicide rates: An assessment of alternative theoretical explanations.* Paper presented at the annual meetings of the American Society of Criminology, Baltimore.

Michelson, W. (1970). *Man and his urban environment.* Reading, MA: Addison-Wesley.

Miethe, T. D., Stafford, M. C., & Long, J. S. (1987). Social differentiation in criminal victimization: A test of routine activities/lifestyle theories. *American Sociological Review, 52,* 184-194.

Miller, E. S. (1981). Crime's threat to land value and neighborhood vitality. In P. J. Brantingham & P. Brantingham (Eds.), *Environmental criminology* (pp. 111-118). Beverly Hills, CA: Sage.

Miller, W. B. (1958). Lower class culture as a generating milieu of gang delinquency. *Journal of Social Issues, 14,* 5-19.

Naroff, J. L., Hellman, D., & Skinner, D. (1980). Estimates of the impact of crime on property values. *Growth and Change, 6,* 121-132.

Rengert, G. F. (1989). Spatial justice and criminal victimization. *Justice Quarterly, 6,* 543-564.

Rosenfeld, R. (1986). Urban crime rates: Effects of inequality, welfare, dependency, region, and race. In J. M. Byrne & R. J. Sampson (Eds.), *The social ecology of crime* (pp. 116-130). New York: Springer.

Rosenfeld, R. (1988, November). *Murder and economic deprivation.* Paper presented at the Conference of the American Society of Criminology, Chicago.

Rosenfeld, R. (1989). Robert Merton's contributions to the sociology of deviance. *Sociological Inquiry, 59,* 453-466.

Rosenfeld, R., & Messner, S. F. (in press). The social sources of homicide in different types of societies. *Sociological Forum.*

Sampson, R. J. (1985). Neighborhood and crime: The structural determinants of personal victimization. *Journal of Research in Crime and Delinquency, 22,* 7-40.

Sampson, R. J. (1986). Neighborhood family structure and the risk of personal victimization. In J. M Byrne & R. J. Sampson (Eds.), *The social ecology of crime* (pp. 25-46). New York: Springer.

Sampson, R. J. (1987). Urban black violence: The effect of male joblessness and family disruption. *American Journal of Sociology, 93,* 348-382.

Sampson, R. J., & Castellano, C. (1982). Economic inequality and personal victimization. *British Journal of Criminology, 22,* 363-385.

Sampson, R. J., & Grove, W. B. (1989). Community structure and crime: Testing social disorganization theory. *American Journal of Sociology, 94,* 774-802.

Sampson, R. J., & Lauritsen, J. L. (1990). Deviant lifestyles, proximity to crime, and the offender-victim link in personal violence. *Journal of Research in Crime and Delinquency, 27,* 110-139.

Savelsberg, J. (1984). Socio-spatial attributes of social problems: The case of deviance and crime. *Population and Environment, 7,* 163-181.

Schuerman, L., & Kobrin, S. (1986). Community careers in crime. In A. J. Reiss & M. Tonry (Eds.), *Communities and crime* (pp. 67-100). Chicago: University of Chicago Press.

Shannon, L. W. (1981). *The relationship of juvenile delinquency and adult crime to the changing ecological structure of the city* (Preliminary Executive Report for National Institute of Justice Grant 79-NI-AX-0081). Iowa City: University of Iowa, Iowa Urban Community Research Center.

Shaw, C. R. (1929). *Delinquency areas: A study of the geographic distribution of school truants, juvenile delinquents and adult offenders in Chicago.* Chicago: University of Chicago Press.

Shaw, C. R., & McKay, H. D. (1942). *Juvenile delinquency and urban areas.* Chicago: University of Chicago Press.

Shaw, M., & McKay, H. (1972). *Juvenile delinquency and urban areas* (2nd ed.). Chicago: University of Chicago Press.

Simcha-Fagan, O., & Schwartz, J. E. (1986). Neighborhood and delinquency: An assessment of contextual effects. *Criminology, 24,* 667-704.

Smith, D. A., & Jarjoura, G. R. (1988). Social structure and criminal victimization. *Journal of Research in Crime and Delinquency, 25,* 27-52.

Smith, D. A., & Jarjoura, G. R. (1989). Household characteristics, neighborhood composition and victimization risk. *Social Forces, 68,* 621-640.

Smith, S. (1986). *Crime, space and society.* Cambridge, UK: Cambridge University Press.

Taub, R. P., Taylor, G., & Dunham, J. (1984). *Paths of neighborhood change.* Chicago: University of Chicago Press.

Taylor, R. B. (1987). Toward an environmental psychology of disorder. In D. Stokols & I. Altman (Eds.), *Handbook of environmental psychology.* New York: John Wiley.

Taylor, R. B. (1989, August). *Developing a longitudinal human micro-ecology of disorder in the urban residential environment.* Presidential address (Division 34) presented at the annual meetings of the American Psychological Association, New Orleans.

Taylor, R. B., & Covington, J. (1988). Neighborhood changes in ecology and violence. *Criminology, 26,* 553-590.

Taylor, R. B., & Covington, J. (1990). Ecological change, changes in violence, and risk prediction. *Journal of Interpersonal Violence, 5,* 164-175.

Taylor, R. B., & Gottfredson, S. D. (1986). Environmental design, crime and prevention: An examination of community dynamics. In A. J. Reiss & M. Tonry (Eds.), *Communities and crime* (pp. 387-416). Chicago: University of Chicago Press.

Williams, K. (1984). Economic sources of homicide: Reestimating the effects of poverty and inequality. *American Sociological Review, 49,* 283-289.

7

Household Coping Strategies and Urban Poverty in a Comparative Perspective

BRYAN R. ROBERTS

THIS CHAPTER TAKES UP the issue of the role of household and family organization in empowering the poor to cope with the challenges of the urban environment. Poor households vary in the degree of success they have in making ends meet in the present or in planning for a better future. Much of that variation is due to factors outside the control of individual households, such as economic cycles or the structure of local job opportunities. Despite these limits on household initiative, the possibility remains open that people can increase their control of their environment through a skillful management of the resources they have/There is considerable debate over the way this increase in control can be achieved. From some perspectives, empowerment is a question of privatizing public concerns by fostering the market participation of the poor, by emphasizing the responsibility of the individual family for the moral and economic welfare of its members, and by freeing individuals and households from the tyranny of state bureaucracy (Douglas, 1989). From other perspectives, household strategies lead to an empowerment based on community strategies of collective action that lessen dependence not only on the state, but also on the market (Friedmann, 1989).[1]

The problem with such perspectives is that they often ignore the fact that poor households cope not in abstract, but with specific environmental challenges, and these vary in time, place, and over the household cycle/The issue, as Hareven (1982, p. 4) points out when discussing family strategies during an earlier period of economic restructuring, is to identify the political and economic contexts under which households can either individually or collect-

AUTHOR'S NOTE: I thank Harley Browning and Omer Galle for helpful comments.

ively exercise some control over their environment, and those under which that control diminishes.

Household strategies, particularly among the poor, have acquired new salience with the economic and political restructuring of the years since the early 1970s.[2] Restructuring imposes special challenges on poor households since one of its main consequences has been to expose individuals more sharply to market forces because of the reduction in public expenditures on urban welfare in such areas as income transfers, collective services, and subsidies of various kinds. As a basic social unit of caring and shelter, the household is peculiarly anchored in urban space and cannot easily move residence in response to changes in the location of job opportunities that restructuring often brings.

Restructuring is, however, only one of several historical transitions that currently affect household coping strategies. We will need to take account of the demographic transition to an aging society, and to smaller family and household units, including single-person ones. There is also the transition toward a service-oriented society in which employment is mainly in the various service occupations, in which the economic participation rates of both males and females are increasingly alike, and in which there is a considerable heterogeneity in employment patterns as a result of the increase in part-time and other "atypical" forms of work (Cordova, 1986).

In this chapter, only a limited number of the contexts affecting the differentiating household strategies can be considered.[3] One of these is the nature of urban spatial organization, including the housing market. The question is whether the current restructuring of urban space facilitates or hinders the creation of the kinds of neighborhood communities that are part of household coping strategies. Another contextual variation is the differences between urban labor markets, both in the range and types of job they offer and in the opportunities they create for the various categories of household members—men and women, young and old.

Finally, we need to consider differences in the organization of welfare, particularly those arising from state welfare policies, such as the types of family and household organization encouraged by public policy, and the extent to which welfare is regarded as a private rather than a public responsibility. Public welfare raises the question of citizenship. In a subsequent section, we will consider whether household strategies can contribute to the practice of citizenship, and whether that practice challenges, or reflects, a top-down definition of rights (Turner, 1990; Van Gunsteren, 1978).

Of the various spheres of citizenship, that of welfare rights has perhaps been the most bitterly contested in recent years, especially in the developed world. Households, especially poor ones, are protagonists as well as victims in that contest. To obtain the resources that are available from a variety of institutions, households must perform tasks and choose among alternatives. Households adjust the mix of their income generating and self-help strategies to take account of what the state has on offer; likewise households join together to claim rights or to provide their own welfare, often becoming the basis of social movements aimed at improving community facilities or gaining a greater share of state resources. Accessing social rights may also have negative implications for the degree of control that households exercise over their environment, by making them dependent clients of bureaucracies.

Both citizenship rights and restructuring are influenced by the particular history of national development. The shape of the postindustrial city and the conditions it provides for households are contingent upon nationally specific patterns of urban development. As Gottdiener (1989) argues, for instance, it would be a mistake to equate the types of urban landscape spawned by the interaction of politics, technological, and economic change in the United States with that of Europe. Furthermore, urban development has been uneven on a world scale, so that the cities of many developing countries are entering their postindustrial phase before the maturing of industrialization or of the urban infrastructure associated with it. Consequently, national differences in social welfare need to be considered as part of the context of household strategies. In some urban contexts, market deficiencies are relatively more severe than those of state provision, in others deficiencies in state provision make the market the only recourse, and in yet others the failure of both state and market to offer basic subsistence makes self-help or nonmonetary reciprocity the main available resource.

To take some account of this variation in patterns of urban development and of the resultant differences in the strategic opportunities and limitations facing households, I will compare three types of urban context. The first corresponds to those European countries with a long history of urbanization and industrialization, where public welfare, though retrenched, is still a significant element in household welfare. My case will be Britain. The second context corresponds to the United States where urbanization and industrialization are more recent, and based on a more decentralized political and economic system, and where welfare has been mainly left to the market and to individual initiative. The third context is that of recently urbanizing and industrializing countries of the underdeveloped world where neither market nor state has been able to provide adequate welfare for the majority of urban inhabitants. My third case will be based on Latin America.

The sociological concern with contemporary household strategies has been uneven in these three cases. It has received most extensive coverage, and over a longer period of time, in the literature on Latin America (Schmink, 1984). It has recently received considerable attention in Europe, particularly the United Kingdom (Crow, 1989). It has received relatively little attention in the United States, apart from the literature on recent immigrants (Aldrich & Waldinger, 1990; Boyd, 1989; Browning & Rodriguez, 1985; Fernández Kelly & García, 1990; Hagan, 1990; Pessar, 1982; Tienda, 1980).[4]

Household strategies are, it will be argued, context bound. Their understanding for policy and other concerns requires both an exploration of their general characteristics, and their specific manifestations under different types of political and economic environment. This chapter thus starts by considering the concept of strategy, and then reviews the various kinds of household strategies. The third subsection begins to examine the question of context by exploring different types of citizenship and the household strategies associated with these types. Then we consider the ways contemporary urban restructuring affects household strategies. Finally, household strategies are contrasted among Latin America, Britain, and the United States.

THE NATURE OF COPING STRATEGIES

Focusing on strategies is a means of providing alternative explanations of social change to those generated by a focus on structural constraints as determinants of behavior. In part, the renewed interest in strategy results from disenchantment with explanations in which structural imperatives associated with processes such as economic modernization or capital accumulation were seen to result in relatively uniform patterns of behavior and organization among subordinate groups, with variation being attributed mainly to the extent and pace of change. Attributing strategies to people, whether as individuals, as households, or as interest groups, signals that despite the importance of structural constraints choice is possible, and that the exercise of choice can result in alternative outcomes.[5]

The concept of household strategy has a similar intent, but needs to be exercised with caution. The over frequent use in the Latin American literature of the term *survival strategies* to describe the ways in which the poor make out in urban and rural areas can create the illusion that the poor have some choice in the matter. There is thus a "myth of survival strategies" (Haguette, 1982) that is little but a euphemism for crushing poverty in which survival depends on selling one's own and one's family labor cheaply and under whatever conditions are offered.

Conversely, too exacting a definition of strategy has its own dangers, particularly in comparative analysis. While it may be useful, in the British or U.S. context, to confine the term *household strategy* to those practices that involve medium- to long-term planning, such as achieving the optimum size of family, purchasing a house, or planning for retirement, this would be less useful in the unstable and, in terms of careers, less predictable urban environments of the developing world. In those environments, even short-term decision making, involving changes in consumption patterns or the labor market allocation of household members, may be critical means of altering the environment in the household's favor.

More analytic flexibility is gained through distinguishing household strategies by whether they are oriented toward coping or toward social mobility. Both types have been identified among urban households, involving, in part, the distinction made above between short-term and longer term strategies and based on whether households and their heads are able or not to pursue stable job careers (Schmink, 1979). Coping strategies can be defined as organizing the household to get by in the short and medium term, while social mobility strategies involve allocative decisions, such as over children's education, the purchase of a house, or improvement in job qualifications, that will bear fruit in the longer term. The two types of strategies cannot always be easily distinguished, especially among the poor. The more formalized the urban environment and the more stabilized the labor market and its rules, the more likely it is that households will be differentiated by whether they have long-term social mobility strategies or shorter term coping strategies.

The general definition of household strategy needs to be a broad one—a set of activities consciously undertaken by one or more members of a household over a period of time, directed toward ensuring the longer term survival of the household unit. Additionally, a household strategy involves calculating between alternative courses of action even though alternatives may differ considerably in the degree of choice they offer.[6] This definition excludes activities undertaken in an ad hoc way, or on a day-to-day basis, to resolve pressing needs or to seize unanticipated opportunities.

Household strategies depend heavily, but not exclusively, on family organization and its prevailing norms. Indeed, the terms *household strategy* and *family strategy* are often used interchangeably, but it is necessary to distinguish them. The household is the unit of co-residence whose members may or may not be kin. Family types, such as nuclear and extended, and the obligations attached to kinship are some of the major variables affecting the household's capacity to implement strategies. Because household members are likely to have different priorities and different conceptions of family welfare, household strategies do not necessarily involve consensus among the various household members.

The range and types of household strategies that are reported in historical and contemporary studies are broadly similar, and will be reviewed in the next section. It is in the particular mix of strategies used by households that the effect of historical time and place appears, whether, for instance, households make intensive local use of their own labor and material resources and community ties or disperse these through labor migration, often as a first step to resettlement as in the case of rural-urban migration (Anderson, 1971; Hareven, 1982; Mingione, 1985).

The mix of alternatives is made more complex because individual household members can pursue them independently or through consensus with other household members. Both male and female heads of household can, for instance, pursue strategies aimed at ensuring household survival that have, at best, only the reluctant consent of other household members. The degree of consensus and of equity in the distribution of household tasks is in fact a further variable in understanding the different types of household strategy as between, for example, male-dominated and joint means of allocating household resources (Jelin, 1984; Redclift, 1988; Rose & Felder, 1988). There is some cross-cultural evidence that among poor households, male-dominated forms of allocation are more common, and can result in women bearing the brunt of reductions in consumption necessitated by economic crisis (Chant, 1985; González de la Rocha, 1986; Vogler, 1990).

Household strategies and the normative constraints surrounding them are shaped by what Hareven (1982) calls the intersection of individual time, family time, and historical time. Household needs and possibilities are inevitably affected by the household cycle as it moves through the stages of formation, consolidation, and finally disintegration. With each stage the balance between dependents and potential wage-earners alters, as do the aspirations of individual household members. As Schmink (1979, pp. 223-236) shows for Brazil, the aspirations of a young couple when they establish a household, and the strategies they can use, will not be the same as those of a household in process of disintegration as the children choose to leave, and the heads of household cease to be economically active.

Historical time is a further factor differentiating the experience of different generations. Each generation is likely to have a specific experience of external opportunities that reflects economic cycles and changes in the structure of job opportunities. Norms, also, are likely to change as in the case, for example, of those concerning married women's paid work or the obligations of young adults to contribute to the family pot. Because new occupations recruit mainly from young age cohorts and declining ones retain older workers, the job opportunities and current work experiences of sons and daughters are likely to be different from those of their parents.

The intersection of these different times resulted in changes in aspiration and strategy among families in Manchester, New Hampshire, in the period from the late 19th century to the early 20th century (Hareven, 1982). These changes and the discontinuities arising from them are at least as great in the modern city, creating further strains on household consensus. Since even neighborhoods, let alone cities, are made up of households at different stages and operating at different historical times, the household cycle is an important source of differentiation of interests and strategies.

The above review suggests important limitations on household strategies as a generalized solution to urban poverty. Household strategies among the poor are likely to be predominantly short-term strategies and to be most prevalent in informal urban environments. These coping strategies are not necessarily based on consensus, and they may depend on an inequitable sharing of tasks. Coping strategies are, indeed, likely to be a source of differentiation between and within households, not as a result of competition over scarce resources, but because of disparities in aspirations resulting from different stages in the household cycle and divergences in the life chances of household members.

TYPES OF STRATEGY

The contemporary urban context, both in developed and developing countries, requires a different mix of strategies than did earlier phases of industrialization and urbanization. Additionally, demographic trends of the contemporary period of urban restructuring provide a general limitation on household strategies.

The mix of strategies reported in the literature can be reduced to combinations of four main types (González de la Rocha, 1988; Mingione, 1987): (a) reducing household expenditures by cutting consumption or ejecting nonproductive members; (b) intensifying the exploitation of internal household resources through self-provisioning and reciprocity with kin and friends; (c) adopting market-oriented strategies, which in the urban context are usually labor market strategies, but not exclusively so as the flourishing of an informal economy indicates; and (d) seeking aid from powerful external agents, such as the state, as of right or in return for political support.

Intensifying the exploitation of internal resources and reducing consumption lessens external dependence, but such strategies are limited by a household's available material and labor resources and by the freedom it has to reallocate these. Strategies aimed at market, state, or other powerful external agencies are less limited in the range of resources they can capture, but they are likely to increase external dependence and may limit the strategic flexibility of households in the future. Reducing expenditures by reducing consumption is a common household coping strategy, but is most likely to be the predominant strategy when households have a substantial margin before basic necessities are reached. In Mexico, where 60% or more of the household expenditures of the poor are on food, increasing labor market participation is the main means of coping with crisis (González de la Rocha, in press). In Britain, in contrast, household members report cutting down on luxuries, not the labor market, as their recourse in times of hardship.[7] Reducing consumption is likely to fall inequitably on household members, particularly, as was noted above, on women. Reducing expenditures in poor households will also usually involve more intensive labor by women and children as time is spent searching out the best buys or preparing foods instead of buying them partly prepared.

These last practices are part of the second type of coping strategy, that of self-provisioning. This is a widely reported household practice, that includes the domestic processing or production of food, making clothes, undertaking repairs, and even the self-construction of housing.[8] In urban economies, however, the third type of strategy—market strategies—is likely to be more important than self-provisioning in generating extra resources.[9] For poor households, these strategies essentially involve maximizing the market return of available household labor—the major economic resource possessed by poor urban households—by finding better paying jobs, placing more members on the labor market or, less significantly, by increasing household production for the market. A study of poor households in the Brazilian city of Fortaleza, which is noted for its large informal economy, found, for example, that 93.7% of household income was generated by paid work (Haguette, 1982).

The state and nongovernmental organizations are frequently treated as resources by poor households, and subject to their strategies, whether individual or collective, to obtain better public facilities or transfers of goods or income.

Urbanization and industrialization have increased household use of externally directed coping strategies by undermining the relatively self-sufficient family economies of both rural and urban areas. Despite the continuing importance of reciprocity and self-provisioning, a relatively small fraction of an urban household's needs can, in the present period, be met solely from its resources or through cooperation with kin or neighbors. Kinship links, for instance, while still serving to reduce external dependence by exchange of aid and goods, are principally used in the modern urban context to provide information and access to external opportunities.[10]

The shift in the balance is due to the range of goods and services that subsistence in the modern city requires—though that range varies from one region to another. Even in times of crisis, individual members of a household may emphasize their own levels of consumption, as much as preserving household and family continuity. Furthermore, many of the necessities of life of the modern city, which are often enforced by government regulation, such as drinking water, sewage disposal, transport, and food, cannot be provided from a household's own resources. Though cooperation among neighbors and kin to secure such goods and services can be effective means of doing so, these strategies, too, depend on the cooperation of external agencies.

Tilly and Scott (1978, p. 212) point out that being a mother becomes a career, as she specializes in dealing with schools and other bureaucratic agencies, or in shopping for the range of goods—shoes, clothing, food, appliances—that are necessities in the modern urban world. The private world of the family is increasingly penetrated by the public world of the market and state regulation (Jelin, 1984, p. 14). Even the activities furthered by self-provisioning and reciprocity often depend heavily on goods and services that originate externally.[11]

It is an open question whether current restructuring shifts further the balance of coping strategies toward external dependence since it includes processes that have conflicting and situationally specific implications for the integrity of the local urban community. In some cases, restructuring means the breakup of established communities, weakening household self-sufficiency in face of market or state. In other cases, as markets stagnate and the state retrenches, it throws households and communities back on their own resources. At best, however, restructuring is likely only to halt the secular trend toward dependence. The modern urban household has, I suggest, only a limited capacity to control its environment through the use of strategies.

A major reason for the decline in the household's significance as a basis for coping with urban life is the changes in the composition of the household unit. This is an important and continuing field of research in both developed and developing countries, and these comments are necessarily tentative. Changes in norms and in economic circumstances make both nuclear and extended families a less typical basis of coping in present times, and weaken the household's external relationships. The decision to establish a new household or to continue as part of an extended family was, in the European past, perhaps the most basic household coping strategy. Giving people more control over this decision, as a result of economic expansion and proletarianization, was an important factor in the timing and pattern of the population expansion of Western Europe (Levine, 1977; Wrigley, 1983).

The subsequent demographic transition in developing and, more hesitantly, in developed countries resulted from a further set of household strategies based on changing urban and labor market conditions, used by both middle and working classes, to raise the quality of life through planning to have fewer children (Alba & Potter, 1982; Seccombe, 1990). Both these strategies were based on the nuclear family as the normal household unit, but making use of wider kin networks as a buffer when the household's own resources were inadequate for survival.

The nuclear family is now less of a norm. Among the poor, the proportion of single-person households and of single-parent families has increased rapidly in the cities of both the developed and developing world, though, as we will see, the pace of change is different in the different contexts. Some of the explanations are demographic—widespread use of contraception, the decline in the proportion of marriages, and the increase in longevity, particularly among women, resulting in single-person households of the elderly.

The socioeconomic changes associated with the modern city are also significant: the greater facility of individuals, particularly adult women, to live on their own or with their children, because of the increase in work opportunities for women, of the availability of public welfare, and of the disincentives to living with unemployed or casually employed males in a situation of greater public tolerance of unwed mothers; and the impact of the media in both developed and developing countries in promoting the ideal of the small family and of the quality of life and consumption associated with it. If to these factors is added the impact of migration and small families in weakening kin networks, the general implication for household strategies is that a greater proportion of household units are likely to be socially isolated in the present as compared with the past.

There are offsetting tendencies, since households may take in kin or non-kin as members to contribute to expenses. Alternatively, members can be ejected to diminish costs. The former practice appears to be more common in the face of crisis, and has a long history in both the developed and developing world. The increase in the numbers of abandoned children in Brazil indicates, however, that the later practice is also a possibility.

The various types of strategies available to poor households in face of these trends depend, for their force, on the continuing importance of the local community. Though this is particularly true for self-provisioning activities and reciprocity, it also applies to labor markets. Even in the formal economy, information networks are likely to be key factors in obtaining work, particularly after unemployment. In the informal economy, which is a local labor market, community relations are essential to job opportunities, whether as casual workers in construction, employees in clandestine workshops, or as home workers in the garment industry.

THE QUESTION OF CITIZENSHIP

The range of strategies discussed above are potentially generally available to households. Which ones predominate and whether they are effective or not depend on a series of contextual factors, among which are the prevailing conceptions of citizenship. Following the distinction developed by T. H. Marshall (1965) between civil, political, and social rights, the sphere of the household in the contemporary urban world is that of social rights.[12] Rights to a basic standard of subsistence, to protection against ill-health, and to adequate care and education for children—the major social welfare rights— are often guaranteed and administered by the state using the household as the unit of management. Turner (1990, p. 209) points out that an important variation in conceptions of citizenship is whether family relations and caring are viewed as essentially private concerns to be resolved within the family, or as appropriate subjects of political activity and organization.

The boundary between private and public definitions of morality is a shifting one, depending on the dominant ideology and subject to change through political activity. The achievement of the welfare state in Britain and in the Scandinavian countries was the result of the political strength of the classes that most needed the public provision of care. Conversely, economic and political restructuring has often involved redefining as private what was previously regarded as appropriate for the public sphere, as when the British conservative government of the 1980s emphasized the "caring capacity of the community" as a substitute for public welfare provision.

Households exercise varying degrees of control over their social rights. Democracy represents a favorable context for gaining and controlling a range of welfare rights—to a minimum family subsistence, to universal education

TABLE 7.1

Types of Social Citizenship

	Autonomous	*Dependent*
Public	Social issue movements over housing, welfare rights, education	Patronage by welfare bureaucracies
Private	Individual social mobility strategies	Social isolation

and health care—but there, too, political centralization combined with low levels of active political participation can result in a situation where rights are defined from above and their beneficiaries have little say in their administration.[13] Bureaucratic administration of needed goods, such as housing or welfare payments, is likely at best to promote a passive acceptance of the technical know-how of those in charge or at worst an individualized hostility, but is unlikely to increase interhousehold cooperation in securing the general welfare. Recipients of benefits may see themselves and be seen by others as state dependents, receiving "handouts" through fixed bureaucratic procedures.

Countering this centralization is the emergence of an increasing number of subgroups with specialized needs that fall outside the purview of the state (Balbo, 1987). A variety of informal networks and voluntary associations have emerged to cater for these needs by providing welfare services or forums for self-expression and mutual aid. The household and its members, particularly women, are key elements in sustaining these activities.

Household strategies are, consequently, critically affected by two major dimensions of contemporary citizenship: (a) the degree of autonomy that individuals feel they exercise over their environment, and (b) the definition of what is a matter of public concern and what is left to the initiative of individuals. Combining these dimensions generates four possible ways in which the prevailing conception of citizenship in a given time and place orientates the strategies of poor households (Table 7.1). These outcomes and the strategies associated with them will often represent different phases of an overall household strategy, as people explore different means of generating the resources they want for themselves and their family, or, alternatively, abandon the effort. With time and shifts in opportunities, this household activity may lead to changes in the prevailing conception of citizenship.[14]

The first type is where households feel they can exercise some control over their future, and seek to advance the family welfare through voluntary organization to improve housing, education, or other urban amenities. This corresponds to a participatory conception of citizenship in which social welfare while defined as a state responsibility is directly and thus locally controlled by its beneficiaries. In the second type, also, households feel capable of planning ahead successfully, but strategies are privatized, aimed at achieving welfare mainly through individual social mobility rather than extending and improving state provision. The correspondence here is with a participatory conception of citizenship in which people participate politically to safeguard their civil rights and their economic interests, but where the welfare of household members is seen as a private family responsibility.

The third and fourth types are marked by low levels of household-based initiative, and correspond to nonparticipatory forms of citizenship. The third type is that where households make use of, and are in frequent contact with, the state or other powerful external agencies, but feel they can exercise little control over the terms of the relationship. The concept of citizenship is a paternalistic one in which welfare rights are recognized, but are defined from above. The fourth type—where households feel socially and politically isolated and have been left to survive through their own devices—corresponds in its extreme manifestation to a denial of citizenship. It is most akin to a situation of social and political marginality.

This typology has some correspondence to the practice and conception of citizenship in the three national cases. Turner (1990) distinguishes between the United States and Britain in terms of the greater emphasis on private solutions to welfare in the former, and more centralized politics in the latter. Thus while the United States fits in general into the autonomous but socially privatized concept of citizenship, Britain uneasily straddles the autonomous and dependent boxes because of its high degree of political centralization. In most underdeveloped countries citizenship in any of its manifestations is not well established. In the case of Latin America, citizenship, though usually enshrined in constitutions, remains—to use Mann's (1987, p. 344) phrase for the situation of France in the 19th century—bitterly disputed. Political rights are frequently withdrawn or subverted, civil rights are often ignored, and social rights are tied to formal employment. Yet the continuing struggle for rights and the survival of democratic forms of government indicate that the countries of Latin America manifest important elements of citizen autonomy.

A major reason that the fit of the typology is inexact at the national level is that citizenship means different things for different subgroups of a population. The social and economic disadvantages of the poor mean that they rarely enjoy fully the rights of citizenship available in their country, and their exclusion can lead to isolation. One consequence that we will note later is that the poor can be relatively more marginalized in the United States or Britain where citizen rights are firmly established than in Latin America where they are not.

HOUSEHOLDS AND RESTRUCTURING

Restructuring has had one major and general effect on households, through its consequences for household income distribution. In both developed and developing countries, the "golden years" of development of relatively rapid growth rates between the 1950s and the 1970s have been replaced by much slower rates of growth subsequently. Though these rates of growth have until recently allowed for an increase in per capita income, this increase has been small compared to previous decades, and distributed unevenly.

In the United States, household income was more concentrated in 1988 than in 1967, with the bottom four deciles of households accounting for 15.1% of income in 1967, and 13.4% in 1988, and the top two deciles accounting for 42.7% in 1967 and 46.3% in 1988(U.S. Bureau of the Census, 1990b, Table 2). A worsening of income concentration also occurred in Britain, increasing the incomes of the top 20% of households by 6% between 1976 and 1985, while decreasing the bottom 20% by 9% (Mellor, 1989). The evidence for Latin America indicates increasing income concentration between 1960 and 1975, reflecting the rapid urbanization of these years, with the top decile absorbing 46.6% of total income in 1960, and 47.3% in 1975. The bottom four deciles of households received 8.7% of income in 1960 and 7.7% in 1975 (Portes, 1985, Table 3). Mexican data, however, show a decline in income concentration between 1977 and 1984, with the bottom four deciles accounting for 11.2% in 1977, and 12.4% in 1984, while the top two deciles reduced their share from 53.7% in 1977 to 51.9% in 1984 (Cortes & Rubalcava, 1990, Cuadro AI.2).

High levels of income concentration and a slow rate of per capita income growth mean that the poorest households, and many middle-income households, are finding it increasingly difficult to make ends meet. The income of a single wage-earner is increasingly insufficient to maintain an urban family at an adequate level of subsistence, making it more imperative for households to place more than one wage earner on the market. The small decline in household income inequality in Mexico, for example, is most likely accounted for by poor households increasing their labor market participation.

In the United States, the "penalty" for being a single-earner household has increased between 1967 and 1988 with the real incomes of single-earner households declining by some 4% in that period. In contrast, the advantage of being a two-earner household has increased, with such households earning 20% more in 1988 than in 1967 (U.S. Bureau of the Census, 1990b, Table 17). Although this trend is partly accounted for by the increasing proportion of single-earner households that are headed by single women, this is itself, as we shall see, part of the problem of household coping strategies among the poor in the United States.

The period of restructuring is, then, the period when social mobility strategies are both less likely to be effective among a wide stratum of urban households and less likely to be perceived by households as a possible option. Restructuring places increasing emphasis on coping strategies, and this is one of the major contrasts with the period up to the mid-1970s when structural changes in urban economies in both developing and developed worlds permitted a considerable amount of occupational mobility. Urban occupational "upgrading" has been reported in these years for the United States (Wright & Martin, 1987), Britain (Mills & Payne, 1989), and also for Latin America (CEPAL, 1989). This upgrading, though still leaving substantial pockets of poverty, did bring higher levels of income to households. It also encouraged increased consumption, particularly of household-related goods and housing amenities, such as new houses, domestic appliances, and leisure equipment.

Contemporary coping strategies, even among the poorest, often include the attempt to maintain these previous levels of consumption.[15] Though restructuring makes it more difficult for poor households to get enough to eat and ensure that there is adequate care for their members, it also calls into question the life-style and type of consumer economy that had developed in the 1970s even among the poorest. Furthermore, the difficulties faced by the poor are intensified by the reduction in the real value of public welfare that has accompanied restructuring, whether in transfer payments or in the quality of the social services.

Apart from its general impact on household incomes, there are, I suggest, three aspects of restructuring that pose particular challenges to poor households, but with different implications in the different national contexts. The first is spatial reorganization, as manufacturing declines in its former bastions, often relocating, as do many office functions, to suburban and small town locations that may be in different regions of the country (Noyelle & Stanback, 1984; Robson, 1988, pp. 1-16). This reorganization does not automatically entail the decay of the central city, since it can become revamped as a center for businesses services, corporate headquarters, entertainment, and gentrified urban living for those with high incomes. Studies in both Britain and the United States suggest, however, that it can result in a certain "ghettoization" of the central city, with increasing income polarization between high-paying jobs in business services and corporate headquarters whose occupants reside in the suburbs, and low-paying jobs in the personal services or sweat-shops or unemployment (Robson, 1988, pp. 25-31; Sassen, 1988). Redevelopment can also break up established low-income neighborhoods to make way for new road and rapid transit systems, high-rise luxury apartments, and office buildings (Soja, 1986).

These processes of urban decay and redevelopment are, it should be remembered, old ones, and not peculiar to restructuring. The impact of restructuring comes from the combination of spatial redevelopment and changes in the urban occupational structure. Areas of the central city not subject to gentrification that traditionally housed low-income families become both socially and economically more isolated than hitherto, with the decline in the number of local job opportunities for which residents are qualified and that offer a living wage. One likely effect, reported for Britain and the United States, is to concentrate urban poverty in particular regions and areas of the city, since jobs are more mobile than the urban poor (Massey & Eggers, 1990; Robson, 1988).

The second major implication of restructuring for the coping strategies of the urban poor is its association with the rise in the proportion of economic activity that does not conform to the standard of stable, full-time employment (A. Marshall, 1987; Roberts, 1989; Standing, 1988). The economically active increasingly include large numbers of part-time workers, casual labor unprotected by labor legislation, and the unemployed. The increasing importance of nonstandard forms of work is based on technological change and the interdependency of the world economy, both of which create incentives for enterprises to use labor more flexibly (Portes & Sassen-Koob, 1987; Robson, 1988, pp. 70-74). The consequences of the rise of nonstandard forms of employment for households are twofold. It makes available a large number of jobs that offer a supplementary income—a wage that is not large enough to maintain a family. Second, it has meant a substantial increase in both developed and developing countries in women's labor market participation, including that of married women with young children (García & Oliveira, in press).

Combined with the relative decline in real value of the single wage noted above, this entails an increase in dual- or multiple-earner families. Avoiding poverty depends on having two heads of household in paid employment. This change in the relation of the household to the labor market, combined with high levels of open unemployment, reinforces the penalties of social and economic isolation. Poor households increasingly need multiple sources of work for their members. This is one of the dynamics behind the increase in small-scale entrepreneurship, through self-employment and small-scale enterprise, and in casual and legally unprotected forms of employment in both developed and developing countries. Both types of income opportunities are often characterized as the rise of the informal economy (Portes & Sassen-Koob, 1987).

Although all of these trends have been reported for Latin American, British, and U.S. cities, they confront the urban poor in these countries in different ways. I develop this argument more fully in the following section, but, to anticipate, the crisis facing the poor in Latin America is the decline in formal employment opportunities in medium and large-scale enterprises, including state agencies, in a context in which such formal employment was the major means of access to public welfare. Urban unemployment rates rose sharply in Latin America in the crisis years of the 1980s to reach a regional average of 8.9% in 1985, but declined to 6.6% by 1987, being particularly severe among the highly educated and the young (International Labour Office, 1989, p. 28). In the absence of public welfare, such as unemployment pay and aid to families below the poverty line, the poor, as we will see, must continue to find a variety of informal income opportunities if they are to survive.

In the United Kingdom unemployment is higher than in Latin America, at 11.9% in 1986. It has a high regional concentration, particularly among the young and in the centers of the decaying industrial cities of the North and Scotland. The welfare state—through unemployment pay and family allowances—is the main source of subsistence for those marginal to the formal labor market. The crisis for poor households is the crisis in the welfare state, since reduced benefits and welfare facilities are likely to entail an increasing social isolation of the very poor and their segregation from other strata of the working class. Small-scale entrepreneurial opportunities and casual employment have increased in Britain in recent years. But as in other parts of Europe, high rates of unionization and government regulation through minimum wages and social security legislation make informal work opportunities much less an alternative source of subsistence than in Latin America or even in the United States (Abrahamson, 1988; Pahl, 1984; Roberts, 1989).

The challenge facing poor urban households in the United States is mainly one of lack of adequate employment—the combination of unemployment and low pay—aggravated by the spatial and ethnic segregation of the poor in central city ghettos. The stagnation in public welfare in recent years, the run-down in public facilities, and the rise in informal income opportunities are also factors affecting the coping strategies of the poor. These factors are not as significant as the decline in employment opportunities because welfare has always been limited in the U.S. context, and the opportunities to generate income from entrepreneurial activities are not, as we will see, widely available to important sections of the poor population.

STRATEGIES
IN COMPARATIVE PERSPECTIVE

I will explore the complex interrelationship between urban context, including prevailing concepts of citizenship, and household strategies by looking at the predominant types of strategy that can be noted during restructuring in the three areas discussed above—Latin America, Britain, and the United States. In each case a brief account of the situation before restructuring is needed to help distinguish those trends due to restructuring from those that are part of the specific pattern of urban development in each area.

LATIN AMERICA

Poor urban households in Latin America have long been reported to be active in the use of economic and political strategies (Nelson, 1979; Schmink, 1984). Household strategies were a necessary accompaniment to the rapid urbanization of the region. The relative decline in economic opportunities in the rural sector, the rapid growth in population, and cities whose economic and social infrastructure were inadequate to meet the needs of the growing urban population, whether migrant or native, meant that households had to make their city, quite literally, by constructing housing, installing their own amenities, and by finding and often inventing income opportunities. All of this was most effectively accomplished through strategically combining household resources and using the family and its contacts as the basis for extending those resources—for example, building up networks of contacts beginning with kin or migrants from the same village, and planning, sometimes over a period of a year or more, the invasion of land.

In the period up to the mid-1970s, social rights acquired an especial saliency for poor households. On the whole, these were years of substantial perceived economic mobility as migrants exchanged rural drudgery and poverty for better income opportunities in the city. What poor households most lacked in the city was an adequate social infrastructure, of which shelter was perhaps the most important component, though schools and health facilities were also demanded. Public welfare provision in Latin America has been minimal in most countries, and social security provision has been tied to employment with key organized sectors of workers, such as public employees, railwaymen, and oil workers, receiving most benefits and obtaining them earliest.

Households had to provide their own welfare and the caring capacity of the community, not state provision, was the main basis for doing so. Household strategies to obtain these social rights quickly converted themselves in many Latin American countries into organized political demands for these rights. Neighborhood-based movements became a significant feature of the Latin American urban landscape in the 1960s and 1970s (Castells, 1983). These social rights movements were often the main channel for popular participation in politics. In contrast, the struggle over economic interests was less generalized and at times excluded the poorest households. Rights were defended by trade unions, but their members did not include many of the urban poor. The trade unions were vertically organized, often co-opted by governments, and only spasmodically exercised political independence. Even in Chile, before the coup of 1972, the neighborhood movement was as important in politics as the trade union movement. Pastrana and Threlfall (1974, pp. 117-127) describe the way, for example, neighborhood organizations and particularly women cooperated with trade unions and lobbied factory workers, during the economic crisis of the last year of Allende, to distribute goods to poor areas of the city.

The situation in Latin American prior to restructuring could thus be described as one in which household strategies were common among the urban poor, both to further economic interests and in terms of securing shelter and other aspects of urban welfare. Whereas household economic strategies rarely resulted in collective organization, whether through labor unions or producer cooperatives, welfare-oriented strategies had become the basis of important though often short-lived and uncoordinated social movements.

With restructuring, two tendencies appear. Among the poor, economic strategies directed toward subsistence become more salient. This trend requires more consensus among household members and places greater strains upon that consensus. Households need more income, but the effort they have to put out to obtain that income increases considerably and the conditions of work worsen. The absence of effective regulation and low levels of subsistence mean, however, that the labor market can absorb increasing numbers of workers, though at lower levels of income. The single wage-earner family, always more of an ideal than a reality for the poor, becomes an increasingly rare basis of subsistence.

Poor households increase substantially the number of their members in the labor market. Household size increases as children delay leaving the parental household or as other kin or as non-kin join the household. Evidence indicates that a greater share of individual income is contributed to household needs. The proportion of single-parent households increases, but the level remains below that reported for the poor in Britain or the United States because of the lack of public welfare and the low pay that women receive compared to men (García & Oliveira, in press; González de la Rocha, Escobar, & Martínez, 1990).[16] Various studies report how, by these means, households managed to offset a decline in subsistence (González de la Rocha, 1988; Benaria, 1989; Selby, Murphy, & Lorenzen, 1990).

The burden of work on particular members increases, however, especially on female heads of household who will frequently both work for pay and carry out all the household chores (Chant, 1985; Redclift, 1988). Women headed some of the most significant protest movements of the late 1970s and 1980s in Buenos Aires, Argentina, organizing against price rises and scarcity, and in defense of human rights (Jelin, 1986). Also, since members have less disposable income to spend on their own purchases of consumer goods, the opportunity for disagreement over spending priorities increases. Countering these individualizing trends is the increasing importance of collective economic strategies. Soup kitchens are one aspect of this, as is the resurgence of what has been called the *barrio* economy (Friedmann, 1989).

The increased salience of economic issues for households has to some extent meant a reduction in the salience of social issues as the priority of the poor becomes that of feeding themselves, and household members have less time to spend on neighborhood organization. González de la Rocha et al. (1990) point out that the crisis in Mexico has tended to privatize household concerns. Also, urbanization has entered a phase of consolidation in which there is less opportunity for invading land and for self-construction. Rental becomes increasingly the predominant form of tenure among the poor, even within squatter settlements, differentiating further the neighborhood-based interests of households.

Since the incomes of most of the working population are low, the main basis of differentiation is neither occupation nor neighborhood, but the household. The major difference made to household income and consumption is the number of economically active members that the household contains and the stage of the household cycle—not whether the household head is formally or informally employed, a non-manual or a manual worker (González de la Rocha et al. 1990, p. 355; Selby et al., 1990).

The poor do not occupy a separate ecological niche within the city, and most neighborhoods are socially heterogeneous. Though there has been a tendency for the professional and managerial classes, government workers and, in some countries, the elite of organized workers, to occupy purpose-built neighborhoods on the outskirts of the city, probably only a quarter of the urban population are segregated in this way.[17] Small-scale enterprises are found throughout poor neighborhoods, and are regarded as welcome sources of employment by poor households, despite the low pay and poor condition of work.

The increasing emphasis on household participation in the labor market makes income opportunities rather than social conditions the major concern of poor households. This is a concern they share with other sectors of the working class and with certain sectors of the middle class. Housing continues to be a concern for the poor and has increased as a concern for the middle classes. Anxieties over prices and incomes creates a basis for temporary political coalitions, but not for organized political movements among a broad spectrum of the urban population. The result is likely to be a move to a participatory but somewhat privatized form of citizenship, emphasizing individual social mobility as much as collective action. In Mexico, for example, political participation through effective voting has increased dramatically during the crisis, while urban social movements are weak.

THE BRITISH CASE

In Britain a stabilized urban working class had been the basis for effective collective action to obtain social rights. Mellor (1989) indicates how British urbanization in the post-World War II period became in important respects a state-directed moral project. Elements in this project were the eradication of the inner-city slums and their replacement by planned neighborhoods in which housing units were rented, under strict guidelines of eligibility, from the state; the fiscal encouragement of suburban housing aimed at the small family; and welfare support for the family unit, both monetary and in terms of social services. The target for extending social rights was the national state, and achieved through the Labour Party representing the interests of the urban working classes.

The urban working classes were important political actors at the local level, controlling city councils and represented on the advisory boards of state welfare agencies. The level of political participation, particularly among the poorest households, was not, however, high. At the local level, working-class politics was controlled by officials of the major national trade unions, and these were guided by national concerns and accommodations as much as by local needs. In this context, the administration of social rights had a distinctly "top-down" character. Households had little say in the allocation of state housing, and family networks were often disrupted by slum clearance. Eligibility for other benefits, such as welfare assistance, also depended on the scrutiny of officials and were not subject to local accommodations.

As in the United States, industrial restructuring and relocation shifted jobs out of the inner cities, resulting in a concentration of unemployment and poverty there, and population immobility (Mellor, 1989). However, due to the spatial concentration of British cities and the numerical importance of state-provided housing—almost 40% of all housing stock in 1982—poor households are less likely to be segregated from other households than is the case in the United States. The major difference with the United States, however, lies in the dependence of the poor on the state. The poor, in Britain, to a much greater extent than in the United States, are the nonworking poor, being either unemployed, infirm, or retired from active work.

Because of the workings of the benefits system and because of the educational and skill similarities of households members, the unemployed are likely to be members of households where all the economically active are out of the labor market. Unemployment pay and family assistance help maintain the integrity of the household, though at levels of income that reduces social contacts (Gallie, Gershuny, & Vogler, in press).[18] Reduction in social services and the low incomes of those dependent on other forms of state assistance (for infirmity or old age) have similar, socially isolating effects. Abrahamson (1988, p. 15) uses the term *withdrawal from society* to describe the equivalent situation of the poor in Copenhagen.

In this situation, household strategies have been most often reported for those households that are in employment and above the poverty line. It is also households above the poverty line that have the highest average number of workers per family. Wives work for pay to increase household consumption of nonbasic goods, unlike the contemporary Latin American situation where such a labor market strategy is most common among the very poor and aimed at maintaining basic consumption.

The result is a certain polarization among the mass of the British working class (Mellor, 1989; Pahl, 1984). On the one hand, there is a prospering working class, using family planning, dual-earner households, and self-provisioning to buy and equip their houses, increase levels of consumption including vacations abroad, and plan retirement. On the other, there is an "underclass," which as Mellor (1989, p. 585) puts it, inhabits the world of making-out. Careful state regulation and welfare provisions reduce the mobility of this class and foreclose informal employment opportunities.

Household coping strategies represent, in this context, a change in the nature of citizenship. Because the poor are highly dependent on the state and because the prospering working class rely on their own economic initiatives and resent the taxation that supports the nonworking poor, welfare is more likely to be seen not as a social right, but as charity handed down from above. The universality of social rights—their acceptance as belonging to the public sphere—is called into question, moving the British case in the direction of that of the United States. Important differences remain, however. Though subverted by contemporary trends, the household, because of public welfare policies, remains a more important element in coping with urban life than it does in the United States. Perhaps more significant is the fact that the universality of social rights in Britain, particularly with respect to health and child care, creates a broad constituency of households ready to defend the welfare state despite the fragmenting impact of the labor market (Gallie & Vogler, 1989). Collective action has been more effective in Britain than in the United States in securing improvements in social and health services and in defending existing services, through mobilizing against hospital closures, support for health workers on strike, and so on. One indication of the continuing high level of acceptance of the principle of universal social rights is that state spending on social, health and educational services has increased both as a proportion of total government spending and as a proportion of the Gross National Product during the years of the Conservative administration—from 1972 to 1987 (World Bank, 1989, Table 11). The increase in spending on the social sector is likely to have been mainly accounted for by the increase in mandatory support (fixed by law) in areas such as health with an aging population and unemployment pay.

THE UNITED STATES CASE

In the United States household strategies have not been noted as a widely used form of coping among the urban poor, and issues such as housing and the provision of public facilities have not given rise to social and political movements to the extent that has been reported for Latin America.[19] The studies that have looked at how the poor make out in a city as active participants have stressed individualized strategies, many of which are based on not using the household as a unit of action. They have also stressed the ethnic dimension—that the struggle is for cultural rights and cultural citizenship, rather than social rights. Studies of the black ghetto, for example, have shown how people cope as age sets, using their own definitions of valued attributes, or by each household member going his or her own way, using the household as a framework into which people come and go but not as a binding constraint (Hannerz, 1969; Liebow, 1967). Although the household, and the social networks based on it, have been shown to be components of coping with life among ethnic groups in poor urban neighborhoods, as in Suttles's (1968) study, they were less salient, for the young at least, than peer groups and the culture of the street.

The controversies over the culture of poverty thesis of Lewis's (1966), and the related analysis of Moynihan's (1965), are illustrative of the perceived lack of strategic behavior among the poor. The very poor, whether black or Puerto Rican, were depicted as immured in a vicious circle of poverty and fatalism that prevented them breaking out of their situation. Critics of this thesis in the United States focused on the structural barriers to social mobility among the poor, particularly those resulting from racial prejudice. Unlike those who attacked the culture of poverty thesis in the Latin American context, no analyst, conservative or liberal, depicted the poor as controlling their environment in any significant way.[20]

The explanation for the absence of collective action among the working poor to achieve social rights, as happened in Britain, is a familiar one. The spatial and economic expansion of the United States, powered by successive waves of immigrants of different ethnic origin, made social mobility an individual and ethnic, rather than a class, phenomenon. Unlike the British case, there was no stable industrial proletariat settled in the same jobs, cities, and neighborhoods from generation to generation creating working-class household identities and forming the basis for a working-class political party. Economic and political decentralization also contributed to the lack of collective struggle at the national level. Social welfare, and such economically relevant rights as that of worker association, is the province of local jurisdictions, making it difficult to provide them through national action.

The exceptions to this pattern of outward spatial and upward social mobility were the Afro-American populations that settled in the major cities from the 1920s onward, and more recently the Hispanic populations. In the post-World War II period, sections of these populations have become increasingly trapped in urban ghettos. The movement of industries from the central cities has shifted jobs away from established inner-city communities. Also, political and economic decentralization has permitted quite rapid shifts in the economic fortunes of regions as industries abandon locations for "green field" sites elsewhere, or as new industries replace old ones but in new locations. There is evidence of labor market polarization in the large cities, as between low-paid, often casual, service jobs, and the highly paid professional, managerial, and technical jobs associated with producer services and corporate headquarters. Intermediate-paying jobs, such as skilled work in industry, have diminished in relative numbers and have often been relocated outside the large cities. An increasing wage spread in most occupations indicates that individual competition for job opportunities has increased, and with greater penalties for those who are immobile (Uchitell, 1990).

The spatial extension of U.S. cities aggravated the isolation of the inner city, creating sharp contrasts between inner-city slums and suburbs and a clearer spatial segregation of different sectors of the working population than occurs in the Latin American cities and even the British cities. This is also true within ethnic groups. There is evidence that the Afro-American middle classes participated in the move to the suburbs from the 1960s onward, a move in part facilitated by civil rights legislation (Wilson, 1987). In the 1970s and 1980s, Wilson (1987: 55-56) argues that the Afro-American poor became increasingly socially isolated in the central cities. The major factor in this isolation appears to be the residential segregation of Afro-Americans, concentrating their poverty and increasing its impact on the local community (Massey, 1990).

They are isolated within their communities since high rates of unemployment, particularly among young males, as is the case in Britain, deprive households of job-related contacts and make stable household formation more difficult. This situation, together with the high percentage of single parent, usually female, headed households, inevitably reduce the opportunities for reciprocity.

While this is not conclusive evidence for the lack of effective household strategies, it does suggest that the opportunities to develop them are less than in the Latin American case. The high proportions of single-parent households among those in poverty in American cities—just over 50% of the total—and the absence of proximate sources of entrepreneurship mean that households have little leeway in developing reciprocity or finding local income opportunities. There is some supporting evidence for this suggestion among the different groups of the Hispanic population. In the Portes and Bach (1985) study, Cuban immigrants were in general able to take advantage of the Cuban economic enclave in the Miami area, using contacts with fellow Cubans to obtain work and housing, while Mexican immigrants were less successful, partly because of a lack of an ethnic economic enclave and being dependent on jobs in Anglo-controlled enterprises.

Mexican-Americans are not as economically marginalized as Afro-Americans since they are mainly concentrated in Southwestern cities where employment opportunities are more numerous than in the Northern cities, though concentrated in blue-collar and casual work and not resulting in significant social mobility from generation to generation (Chapa, 1988). The Los Angeles labor market for Mexican-Americans and recent Mexican immigrants is similar in important respects to that of Mexican cities, with a large number of informal employment opportunities as well as low paid blue-collar work. Kinship and community ties are significant means of accessing these opportunities and providing a generalized source of social and economic support (Massey, Alarcón, & Gonzaléz, 1987, pp. 253-284). The household unit remains, then, a more significant basis for coping among the Hispanic population than among the Afro-American population, but with differences among the Hispanic population that reflect differences in their urban situation such as the concentration of Puerto Ricans in the economic ghettos of the North-East, of Cubans in the Miami "enclave," and of Mexican-Americans in the Southwest. The proportion of female-headed households among Afro-Americans is greater than that among Hispanics, both among all households (42.4% to 20.8% in 1989), and among households in poverty (73.5% to 43.9%) (U.S. Bureau of the Census, 1990a, Table 20). Among Puerto Ricans, however, the proportion of female-headed households in 1980 among all Puerto Rican families was 36.5%, compared to 18.9% for the Mexican origin population, and 16.0% for the Cuban origin population (Bean & Tienda, 1987, Table 6.8).

The overwhelming problem of poor households in the United States is social and spatial isolation, created by ghettoization and a lack of employment opportunities. It has resulted in clear ethnic differences in poverty and in the capacity of the household to mitigate poverty through coping strategies. In this situation, citizenship in its social dimensions is denied to a substantial minority of Americans, while the issue of cultural citizenship has become more salient through marginalizing certain ethnic groups from the consumption and employment opportunities of American society.

CONCLUSION

Household coping strategies make clear some of the basic issues in urban poverty and the policies needed to prevent it. There is little evidence, even in the Latin American case, that poor households can solve their problems through a more skillful and determined use of their own resources. In the

Mexican case, for example, the poor have used the household to avert the worst consequences of economic crisis, but their situation in terms of levels of consumption is precarious, and at best means that previously improving social indicators, such as infant mortality, will not worsen. The poor have borne the major part of the economic burden of the crisis, working more hours for less pay and with less subsidy in order that, through fiscal austerity and a favorable investment climate, the economy will recuperate.

There is, however, no clear way to remedy poverty by external intervention. Abrahamson (1988) contrasts three approaches to urban poverty—advocacy, the public policy approach, and mobilization. *Advocacy* has been more common in the United States where groups have organized to draw attention to the plight of the poor, and argue on their behalf. *Public policy* has characterized most European countries where the power of organized labor and of political parties representing the working classes have forced concessions from economic elites. Public policy in these countries is often based on corporatist politics in which employers and labor organizations agree on development priorities. *Mobilization* is based on the poor organizing themselves to control their immediate environment and to demand redistribution and a better share of employment opportunities.

In Chile, these three approaches competed with each other in the politically turbulent years of the governments of Eduardo Frei and Salvador Allende from 1964-1973 (Pastrana & Threlfall, 1974, pp. 66-74). Each approach to demarginalizing poor households had limitations, apart from the final one imposed by the military coup of 1973. The first approach was mainly used by the Christian Democrats and was based on the state as a source of assistance to the poor, providing them with material help and encouraging them to organize themselves to improve housing and amenities. This approach encouraged clientelism, and by creating brokers within the communities further differentiated them and diminished participation. The second was favored by some elements in the Popular Unity government of Allende and was based on vertical representation through trade unions and other centralized popular organizations. This corporatist model allowed little effective grass-roots participation, and become increasingly ineffective as trade unions were weakened by the economic crisis. The third was associated with the more left-wing parties of the Popular Unity government and based on local-level participation in controlling and developing different aspects of neighborhood life. Even in this case, however, local participation was limited, partly because of the tight control exercised by political parties but mainly by the individualizing effects of the economic crisis on household strategies (Castells, 1983).

Even in the Chilean case, the neighborhood was and continues to be a fragile basis for solidarity in the face of the state and the market economy. In the 1980s, with high levels of unemployment, neighborhood-based collective action in Chile remains fragmented as each household seeks its individual survival (Tironi, 1987).

Neither external initiatives nor the solidarities engendered by household and community are likely alone to be the basis for effective collective action to remedy poverty. In the evolution of citizenship we can, in contrast, see the possibilities of change. These depend, however, on urban context. In Latin America, the advancement of political citizenship provides the current force for equity in society. In Britain, social citizenship remains a powerful rallying point for political coalitions at a time when economic interests and the politics based on these have declined as a source of class allegiance. In the United States, the strongest current force for change appears to be the issue of cultural citizenship that holds out the possibility of coalitions broad enough to effect the policy changes required to eliminate poverty.

NOTES

1. Abrahamson (1988) points out that the congruence between left and right political positions on empowering the poor through encouraging the household and family unit to take more responsibility for its own welfare is due to two contradictory reasons: the financial strain of the growth in public services and criticisms of the impersonal workings of public benefits and services.

2. In a later section, I will consider in more detail what constitutes strategy, and the issue of the degree of consensus that the concept of household strategy implies.

3. These contexts can be conceived as opportunity structures, such, for example, as the range of housing, job or welfare alternatives among which household members can realistically choose in a particular city.

4. There is, of course, a considerable body of literature on urban poverty in the United States, but this has mainly been concerned with either the cultural or the structural limitations that the poor face (Lewis, 1966; Moynihan, 1965; Piven and Cloward, 1971). When attention is given to how the poor cope with their environment, the focus, as Jelin (1984) points out in comparison with Latin American studies, is on individuals and their networks, rather than household strategies.

5. Mann (1987), for example, uses the concept of "ruling class strategies" to argue against the notion that citizenship must necessarily follow a predetermined evolution: ruling classes in Britain, France and Germany defined citizenship in different ways in resolving the dilemma of how to commit subordinate classes to the nation, while still retaining control and the lion's share of the benefits.

6. See Crow's (1989) discussion of the uses of the term strategy. In which strategy corresponds to Weber's definition of calculative rationality, and can be contrasted with other forms of rationality, such as affective and traditional.

7. These findings, and others that will be cited, are drawn from the preliminary analyses of researchers belonging to the Economic Change and Social Life research program of the British

Economic and Social Research Council. The study will be reported in a series of volumes to be published by Oxford University Press, and further information can be obtained from the coordinator, Duncan Gallie, of Nuffield College, Oxford.

8. These strategies have been widely reported in Europe, the United States and Latin America, and included spatial mobility to balance available income, the need for space, and proximity to employment opportunities (Schmink, 1979; 249-279). Even the self-construction of housing, a strategy that is highly time and resource consuming, is almost as widely used in a highly industrialized country such as Germany as in Latin America (Gtatzer and Berger, 1988: 517).

9. Warman (1985) reports the changing subsistence strategies of Mayan peasants, and argues that self-provisioning strategies are no longer widely used, despite a long tradition of families weaving their own clothes, making their shoes, etc. Women's time is spent in domestic out-work, such as making hammocks for the tourist industry. In Socialist economies, also, market strategies—through the informal economy—rather than self-provisioning or seeking more state aid appear to be the major means open to households to improve their control of their environment (Stark, 1989).

10. Urban household studies consistently report that exchange of material aid among kin or friends is far less frequent than is exchange of information, help with obtaining a job etc.

11. Self-provisioning in the contemporary urban household uses sophisticated equipment, whether it be for food-processing, decorating, or other do-it-yourself activities. Gershuny (1988) relies on this to make his argument that technological innovation enables households to self-provision in innovative ways, creating a demand for manufactures, and reducing unpaid domestic work time.

12. Civil and political rights are, in contrast, mainly individual rights, and their exercise, whether in terms of the rights of women, of children, or of individual wage-earners, are less constrained by household considerations, and may, at times, run counter to the integrity of the household unit.

13. Turner (1990) categorizes Britain in the "rights from above" category, because of the way elites gradually handed down political and social rights, contrasting it on this dimension with France and the United States where popular struggle played a greater part historically in defining rights.

14. This typology is essentially the same as that of Turner (1990), with the autonomy/dependence dimension corresponding to his dimension of whether or not the source of rights comes from above or below. My focus on contemporary household strategies makes the perspective, however, a different one in that above/below, in Turner's typology, corresponds to the historical origins (handed down by elites or gained by popular struggle) of the rights, while my use of autonomy/dependence refers to whether or not people further their rights through voluntary activity of various kinds.

15. Lourdes Benaria (1989) reports that among the very poorest households in Mexico City—those she categorizes as living in extreme poverty— 85% had a TV, 55% a refrigerator, 65% a large stove, 40% a washing machine and 45% a sewing machine. She notes that the crisis is likely to have made some of these items useless for lack of repair.

16. Tilly and Scott (1978: 121) quote a comment from 19th century Paris that illustrates the same dilemma: "If there is one opinion that is widespread among popular classes that a single woman cannot earn a living in Paris. . . . alternative to live in privation or to marry."

17. Portes (1989) reviews the diverse outcomes of the current economic crisis in terms of urban spatial organization: in some cities, the tendency for the middle classes to move to suburbs has become more pronounced, but, in others, the impoverishment of the middle classes have led them to seek the cheaper land and housing of poor neighborhoods.

18. Pahl (1984) describes how the elderly and unemployed in his sample of households could *least* make use of self-provisioning strategies since they could not easily reciprocate the help of others, nor could they afford the equipment or material needed.

19. Interest in household strategies in the United States has tended to focus on what, in Latin American terms, can only be described as middle-class strategies—the extent to which households self-provision through activities such as mowing the lawn, decorating and other do-it-yourself activities.

20. An exception is Stack (1974). Critics of Lewis's (1966) work in the Latin American context also noted the structural factors producing poverty, but they also showed the activism of the poor in managing their economic and political environment (Nelson, 1979; Lomnitz, 1977). Also, whereas conservative viewpoints in the United States have tended to attribute ghetto poverty to cultural factors and stressed more effective policing as remedies, conservatives in Latin America have tended to argue that the poor—through the informal economy—already do a good job of coping, and one that would be improved if they were less exposed to bureaucratic regulation.

REFERENCES

Abrahamson, P. (1988). *Social movements and the welfare state: Comparing the struggle against urban poverty in Scandinavia and the U.S.* (Arbejdspapir NR.5-1988). Copenhagen: Sociologisk Institut.

Alba, F., & Potter, J. (1982). Population and development in Mexico since the 1940s: An interpretation. *Population and Development Review, 12*(1), 47-73.

Aldrich, J. E., & Waldinger, R. (1990). Ethnicity and entrepreneurship. *Annual Review of Sociology, 16,* 111-135.

Anderson, M. (1971). *Family structure in nineteenth century Lancashire.* Cambridge, UK: Cambridge University Press.

Balbo, L. (1987). Family, women, and the state: Notes toward a typology of family roles and public intervention. In C. S. Maier (Ed.), *Changing boundaries of the political.* (pp. 201-219.) New York: Cambridge University Press.

Bean, F., & Tienda, M. (1987). *The Hispanic population of the United States.* New York: Russell Sage.

Benaria, L. (1989, November). The Mexican debt crisis: Restructuring the economy and the household. Paper presented at the ILO Workshop on Labour Market Issues and Structural Adjustment, Geneva.

Boyd, M. (1989). Household and family in immigration. *International Migration Review, 23,* 638-670.

Browning, H. L., & Rodriguez, N. (1985). The migration of Mexican indocumentados as a settlement process: Implications for work. In G. J. Borjas & M. Tienda (Eds.), *Hispanics in the US economy.* Orlando, FL: Academic Press.

Castells, M. (1983). *The city and the grassroots.* London: Edward Arnold.

CEPAL. (1989). *Transformaciones ocupacionales y crisis en America Latina.* Santiago, Chile: CEPAL.

Chant, S. (1985). Family formation and female roles in Queretaro, Mexico. *Bulletin of Latin American Research, 4,* 17-32.

Chapa, J. (1988). *Are Chicanos assimilating* (Working Paper 88-8). Berkeley: University of California, Institute of Government Studies.

Cordova, E. (1986). From full-time wage employment to atypical employment: A major shift in the evolution of labor relations. *International Labor Review, 125*(6), 641-658.

Cortes, F., & Rubalcava, R. M. (1990). *Equidad via reducción la distibución del ingreso en México (1977-1984). Mimeo. Centro de Estudios Sociologicos, El Colegio de Mexico, Mexico, D.F.*

Crow, G. (1989). The use of the concept of "strategy" in recent sociological literature. *Sociology, 23*(1), 1-24.

Douglas, J. (1989). *The myth of the welfare state.* New Brunswick, NJ: Transaction Books.

Fernández Kelly, M. P., & García, A. M. (1990). Power surrendered, power restored, The politics of home and work among Hispanic women in southern Florida. In L. Tilly & P. Guerin (Eds.), *Women, Politics, and Change.* New York: Russell Sage.

Friedmann, J. (1989). The dialectic of reason. *International Journal of Urban and Regional Research, 13,* 217-236.

Gershuny, J. (1988). Time, technology and the informal economy. In R. E. Pahl (Ed.), *On work: Historical, comparative and theoretical approaches* (pp. 579-597). Oxford, UK: Basil Blackwell.

Gallie, D., & Vogler, C. (1989). *Labour market deprivation, welfare and collectivism* (Working Paper No. 15). Swindon, UK: Social Change and Economic Life Initiative, ESRC.

Gallie, D., Gershuny, J., & Vogler, C. (in press). Unemployment, the household and social networks. Mimeo chapter for volume on Unemployment of Social Change and Economic Life Initiative of the British Economic and Social Research Council. London: Oxford University Press.

García, B. & Oliveira, O. de. (in press) Cambios en la presencia femenina en el mercado de trabajo: 1976-1987. *Demografia,* El Colegio de Mexico.

Glatzer, W., & Berger, R. (1988). Household composition, social networks and household production in Germany. In R. E. Pahl (Ed.), *On work: Historical, comparative and theoretical approaches* (pp. 513-526). Oxford, UK: Basil Blackwell.

González de la Rocha, M. (1986). *Los recursos de la probreza: Familias de bajos ingresos de Guadalajara.* Guadalajara, Mexico: CIESAS, El Colegio de Jalisco.

González de la Rocha, M. (1988). Economic crisis, domestic reorganisation and women's work in Guadalajara, Mexico. *Bulletin of Latin American Research, 7*(2), 207-223.

González de la Rocha, M. (in press). Crisis, food consumption and access to services: The Guadalajara working class. In A. Escobar & M. Gonzalez (Eds.), *The Mexican crisis of the 1980s: State action, social impacts and social responses.* San Diego, CA: Center of US-Mexican Studies.

González de la Rocha, M., Escobar, A., & Martínez, M. (1990). Estrategias versus conflicto. Reflexiones para el estudio del grupo doméstico en época de crisis. In G. de la Peña et al. (Eds.), *Crisis, conflicto y sobrevivencia* (pp. 351-368). Guadalajara, Mexico: Universidad de Guadalajara/CIESAS.

Gottdiener, M. (1989). Crisis theory and socio-spatial restructuring: The US case. In M. Gottdiener & N. Komninos (Eds.), *Capitalist development and crisis theory: Accumulation, regulation and spatial restructuring* (pp. 365-390). New York: St. Martin's.

Hagan, J. M. (1990). *The legalization experience of a Mayan community in Houston.* Doctoral Dissertation, University of Texas at Austin.

Haguette, T. (1982). *O mito das estragégias de sobrevivencia.* Fortaleza, Brazil: Universidade Federal do Ceará.

Hannerz, U. (1969). *Soulside: Inquiries into ghetto culture and community.* New York: Colombia University Press.

Hareven, T. (1982) *Family time and industrial time: The relationship between the family and work in a New England industrial community.* Cambridge, UK: Cambridge University Press.

International Labour Office. (1989). *World Employment Report.* Geneva: ILO.

Jelin, E. (1984). Familia y unidad domestica: Mundo publico y vida privada. *Estudios Cedes,* Buenos Aires.

Jelin, E.(comp.), (1986). *Ciudadania e identidad: La mujer en los movimientos sociales en America Latina.* Geneva: UNRISD.

Levine, D. C. (1977). *Family formation in an age of nascent capitalism.* New York: Academic Press.

Lewis, O. (1966). *La vida: A Puerto Rican family in the culture of poverty.* New York: Random House.

Liebow, E. (1967). *Tally's corner.* Boston: Little, Brown.

Lomnitz, L. (1977). *Networks and marginality: Life in a Mexican shantytown.* New York: Academic Press.

Mann, M. (1987). Ruling class strategies and citizenship. *Sociology, 21*(3), 339-354.

Marshall, A. (1987). *Non-standard employment practices in Latin America.* Discussion Paper, DP/6/87. Geneva: International Institute for Labor Studies.

Marshall, T. H. (1965). *Social policy in the twentieth century.* London: Hutchinson.

Massey, D. (1990). American apartheid: Segregation and the making of the underclass. *American Journal of Sociology, 96,* 329-357.

Massey, D., Alarcón, R., Durand, J., & González, H. (1987). *Return to Aztlan.* Berkeley: University of California Press.

Massey, D., & Eggers, M. (1990). The ecology of inequality: Minorities and the concentration of poverty, 1970-1980. *American Journal of Sociology, 95,* 1153-1188.

Mellor, R. (1989). Transitions in urbanization: Twentieth-century Britain. *International Journal of Urban and Regional Research, 13*(4), 573-596.

Mills, C., & Payne, C. (1989). *Service class entry in worklife perspective* (Working Paper 10). Social Change and Economic Life Initiative, ESRC, Swindon, UK.

Mingione, E. (1985). Social reproduction of the surplus labor force: The case of Southern Italy. In N. Redclift & E. Mingione (Eds.), *Beyond Employment: Household, gender and subsistence* (pp. 14-55). Oxford, UK: Basic Blackwell.

Mingione, E. (1987). Urban survival strategies, family structure and informal practices. In M. P. Smith & J. R. Feagin (Eds.), *The capitalist city: Global restructuring and community politics* (pp. 297-322). Oxford: Basic Blackwell.

Moynihan, D. P. (1965). *The Negro family: The case for national action.* Washington, DC: U.S. Department of Labor.

Nelson, J. (1979). *Access to power: Politics and the urban poor in developing countries.* Princeton, NJ: Princeton University Press.

Noyelle, T., & Stanback, T. M. (1984). *The economic transformation of American cities.* Totowa, NJ: Towman & Allanheld.

Pahl, R. E. (1984). *Divisions of labour.* Oxford: Basic Blackwell

Pastrana, E., & Threlfall, M. (1974). *Pan, techo y poder: El movimiento de pobladores en Chile (1970-1973).* Buenos Aires: Ediciones Siap/Planteos.

Pessar, P. A. (1982). The role of households in international migration and the case of the US-bound migration from the Dominican Republic. *International Migration Review, 16*(2) 342-364.

Piven, F. F., & Cloward, R. A. (1971). *Regulating for poor: The functions of public welfare.* New York: Academic Press.

Portes, A. (1989). Latin American urbanization during the years of crisis. *Latin American Research Review, 24,* 7-44.

Portes, A. (1989). Latin American class structures. *Latin American Research Review, 20,* 7-39.

Portes, A., & Bach, R. L. (1985). *Latin journey: Cuban and Mexican immigrants in the United States.* Berkeley: University of California Press.

Portes, A., & Sassen-Koob, S. (1987). Making it underground. *American Journal of Sociology, 93,* 30-61.

Redclift, N. (1988). Gender, accumulation and the labour process. In R. E. Pahl (Ed.), *On Work: Historical, comparative and theoretical approaches.* (pp. 428-448). Oxford, UK: Basil Blackwell.

Roberts, B. R. (1989). The other working class: Uncommitted labor in Britain, Spain and Mexico. In M. L. Koln (Ed.), *Cross-national research in sociology* (pp. 352-372). Newbury Park, CA: Sage.

Robson, B. (1988). *Those inner cities.* Oxford, UK: Clarendon Press.

Rose, M., & Felder, S. (1988). *The principle of equity and the labour market behaviour of dual earner couples* (Working Paper 3). Social Change and Economic Life Initiative, Economic and Social Research Council, Swindon, UK.

Sassen, S. (1988). *The mobility of labor and capital: A study in international investment and labor flow.* Cambridge, UK: Cambridge University Press.

Schmink, M. (1979). Community in ascendance: Urban industrial growth and household income strategies in Belo Horizonte, Brazil. Doctoral dissertation, The University of Texas at Austin.

Schmink, M. (1984). Household economic strategies: Review and research agenda. *Latin American Research Review,* 87-101.

Seccombe, W. (1990). Working-class fertility decline in Britain. *Past and Present, 126,* 151-188.

Selby, H., Murphy, A. D., & Lorenzen, S. A. (1990). *The Mexican urban family: Organizing for self-defense.* Austin: University of Texas Press.

Soja, E. (1986). Taking Los Angeles apart. *Environment and Planning D,* 255-272.

Stack, C. B. (1974). *All our kin: Strategies for survival in a black community.* New York: Harper & Row.

Standing, G. (1988). *European unemployment, insecurity and flexibility: A social dividend solution* (World Employment Programme Labour Market Analysis Working Paper No. 23). Geneva: ILO.

Stark, D. (1989). Bending the bars of the iron cage: Bureaucratization and informalization in capitalism and socialism. *Sociological Forum, 4* (4), 637-664.

Suttles, G. D. (1968). *The social order of the slum.* Chicago: University of Chicago Press.

Tienda, M. (1980). Familialism and structural assimilation of Mexican immigrants in the United States. *International Migration Review, 14,* 383-408.

Tilly, L., & Scott, J. (1978). *Women, work, and family.* New York: Holt, Rinehart & Winston.

Tironi, E. (1987). Pobladores e integración social. *Proposiciones* (Santiago de Chile) *14,* 64-84.

Turner, B. S. (1990). Outline of a theory of citizenship. *Sociology, 24*(2), 189-217.

Uchitell, L. (1990, August 14). Unequal pay widespread U.S. *The New York Times,* p. C1.

U.S. Bureau of the Census. (1990a). *Money income and poverty status in the United States, 1989* (Current Population Reports, Series P. 60, No. 167). Washington, DC: Government Printing Office.

U.S. Bureau of the Census. (1990b). *Trends in income by selected characteristics: 1947-1988* (Current Population Reports, Series P. 60, No. 167). Washington, DC: Government Printing Office.

Van Gunsteren, H. (1978). Notes of a theory of citizenship. In P. Birnbaum, J. Lively, & G. Parry (Eds.), *Democracy, consensus and social contract* (pp. 9-35). London: Sage.

Vogler, C. (1990). *Labour market change and patterns of financial allocation within households.* (Social Change and Economic Life Initiative Working Paper). ESRC, Nuffield College, Oxford.

Warman, A. (1985). Estrategias de sobrevivencia de los campesinos Mayas. *Cuadernos de Investigación Social,* 13. Mexico, D. F.: Instituto de Investigaciones Sociales, UNAM.

Wilson, W. J. (1987). *The truly disadvantaged: The inner city, the underclass, and public policy.* Chicago: University of Chicago Press.

World Bank. (1989). *World development report 1989.* New York: Oxford University Press/World Bank.

Wright, E. O., & Martin, D. (1987). The transformation of the American class structure. *American Journal of Sociology, 93,* 1-29.

Wrigley, E. A. (1983). The growth of population in eighteenth-century England: A conundrum resolved. *Past and Present, 98,* 121-150.

8

The Privatization of Collective Consumption

RAY FORREST

THIS CHAPTER EXPLORES the development of state policies designed to privatize areas of collective consumption. It is set within a general theoretical context that emphasizes a shift in debates from examining the consequences of the expansion of collective welfare provision in capitalist societies to a concern with the limits to its contraction. The chapter explores initially the relevant theoretical context and the extent to which the rhetoric of privatization has matched the reality. The terminological ambiguity of privatization and collective consumption are also examined briefly. The chapter provides explanations of why privatization has occurred and the different forms it has taken. In particular it emphasizes that different regimes of collective provision provide the preconditions for different forms of privatization. While the chapter is wide-ranging and cross-national in scope, it focuses particularly on the United Kingdom and the United States and the general discussion is grounded in an analysis of developments in housing provision. This focus provides a contrast between two systems that have very different histories of state intervention in housing provision but apparently similar privatization policies.

COLLECTIVE CONSUMPTION AND PRIVATIZATION: THE CONCEPTS IN CONTEXT

In much of the contemporary academic literature, collective consumption is used as a shorthand for what Marxists once said about state provision and that proved to be false. Leaving aside for the moment the need to differentiate forms of state provision and the taxonomic problems of such an all-encompassing label, the original formulation is assumed to be roughly as follows. In a capitalist society the State intervenes to reproduce necessary labor power; this intervention takes the form empirically of collective provision such as state education, state health services, state housing, public parks, and various social services; these forms of provision will continue to expand in order to maintain the general conditions for capital accumulation; this expansion of state provision undermines profitability but a contraction of those elements of collective consumption would undermine the social and economic fabric of advanced capitalist societies. Hence there is a basic contradiction; ultimately capitalism can neither live with nor live without an expanding sector of goods and services collectively provided.

This conception of developments in the sphere of social consumption has been criticized along two main lines. First, whereas the expansion of collective consumption is what the theory would suggest, the opposite seems to have occurred. As expressed by Saunders and Harris (1990): "Back in 1979 we believed that the state would become ever more entangled in economic and social provisioning. What has actually happened is that it has privatized where theories led us to expect that it would socialise" (p. 58). And second, the contraction of state provision that has occurred through measures such as budgetary cuts, changes in eligibility criteria, asset sales, and deregulation has not apparently produced the political, social, and economic impacts that one might expect. Flynn (1988) commenting on the British situation observes the puzzling paradox that "despite major reductions in public housing expenditure and a significant deterioration of housing opportunities for a substantial section of the population, there has not been any widespread discontent or organised protest" (p. 289). Where are the urban movements mobilizing to resist the cuts? Where is the evidence of widespread resistance to the dismantling of the postwar welfare state?

Certainly the overt discontent of the early 1980s was given greater significance than was subsequently justified. Writing in 1984, Harloe and Paris suggested that political mobilization could become more extensive if further major cuts in collective provision occurred.

> Attempts to decollectivize consumption that are pushed beyond a certain point and in certain directions may contribute to this process of conflict and disintegration within the conventional political system. The reality or even the fear of this occurring is likely to be a real constraint on government action. (p. 92)

In 1990, however, there is no real evidence of major conflict or disintegration despite a decade of apparent fiscal retrenchment. There is evidence of impoverishment of minorities and greater social division, but it would be difficult to justify the view that the threat of social unrest has acted as a brake in the political project of Thatcherism, although pragmatic electoral considerations have certainly played a part. In the United Kingdom the urban riots of 1981 were initially hailed as the first sign of a significant political backlash to decollectivization and disinvestment in the urban structure. More widespread and long-lasting discontent did not, however, emerge, and the riots were reinterpreted as a response to more repressive policing measures with a strong racial dimension. They certainly failed to occasion more widespread discontent in subsequent years when on the surface at least social and economic conditions in working-class urban areas had deteriorated.

We are left therefore with something of a conundrum. Critiques of previous conceptions of the role of collective consumption assume, however, a severity and immediacy of cuts that may not have occurred. It may be that the rhetoric of privatization has been out of step with the reality and that the restructuring and reduction of collective consumption have been less extensive or at least more selective than was anticipated. One possible reason for the lack of reaction therefore is that there has been no throughgoing root and branch transformation of welfare provision and that the demise of collective consumption has been overstated. Another possibility is that the consequences of policy changes may not become apparent for some time. For example, disinvestment in low-income housing does not have immediate effects. Homelessness and affordability problems are not created overnight but over a number of years. Services such as public transport or medical care generally deteriorate over some time as opposed to changes in, say, eligibility for state benefits, which have an immediate impact. Some sections of the working class may have borne the brunt of sharp cuts in state benefits but the mass of the population is more likely to experience the slower and more cumulative corrosive effects of disinvestment.

There is also a further complication in assessing the recent history of welfare restructuring. The issue may not be as simple as whether or not there have been substantial cuts or more modest disinvestment. It may be more appropriate to view developments in terms of reshaping and reorientation of state provision rather than a simple withdrawal. Implicit in some of the initial conceptions of collective consumption was a rather one-dimensional view of state provision and its relation to capital accumulation. This is well expressed by Szelenyi (1981), who argues that structuralist Marxism "overemphasized the self-perpetuating nature of capitalism in the capitalist core countries and underestimated the possibilities of alternatives within these societies" (p. 3).

One dimension of these alternatives was in the development of different forms of welfare states and in conceptions of welfare that encompassed the domestic and informal spheres. For Szelenyi (1981) a comparative and global perspective was required and one that emphasized the dynamic nature of capitalism and the shifting relationships involved. These shifts in the sphere of collective consumption could refer to a reshaping of the relations between state and market forms, a more complex social division of state subsidy and support rather than any simple reduction of state activity along a single continuum. Such a conception of the state's involvement in welfare provision has in fact a long pedigree in that now rather anachronistic British discipline of social administration. Writers such as Titmuss (1976) recognized long ago that state support of social consumption took various forms and the legitimacy and status of those forms was itself a matter for class analysis. Forms of subsidy that benefit a predominantly middle-class population (e.g., mortgage interest tax relief, student grants) are regarded as socially legitimate while offers targeted on the working class (e.g., subsidies for state housing, social security) are condemned as profligate and wasteful. More recently Harrison (1986) has connected those debates on the social division of welfare with conceptions of collective consumption and argued, inter alia, that "there is no case for drawing an arbitrary boundary between visible and less visible parts of the welfare state, or between public and private sectors, when considering how cities have developed" (p. 234). Thus in discussing recent developments in relation to collective consumption we have to work with a broad conception of state intervention.

The points of contact between the original formulations of Castells (1976) on collective consumption and contemporary debates on privatization are extremely diverse. Complications arise in relation to definitions of both collective consumption and privatization. While both terms convey a general tendency they are no more than useful shorthand for processes that have taken different forms in different nation states at different times. Castells himself, contrary to some interpretations, recognized the contingent nature of the processes he was discussing. In the Afterword to *The Urban Question* he states:

> This collective consumption is, therefore, consumption of commodities whose production is not assured by capital, not because of some intrinsic quality, but because of the specific and general interests of capital; thus the same product (housing, for example) will be treated both by the market and the state, and

will therefore be alternately a product of individual or collective consumption, according to the criteria, which will change according to the historical situation. (p. 460)

So there is no presumption that, for example, direct state provision of housing will necessarily expand. Different historical circumstances will impact on the pattern and mode of provision in specific ways.

The quotation above also contains a confusion that a number of analysts have picked up (e.g., Pahl, 1977) namely the distinction between individual and collective consumption. In the example used by Castells (1976)—housing—one of the ways in which state housing provision differs from the provision of public parks is that generally the use of a dwelling is not shared beyond an individual or household. But in this respect there is no difference between state housing and home ownership. Indeed, collective consumption interpreted in this way would be more closely associated with the most commodified form of housing provision, private renting in which multiple occupation is more common. The confusion arises from a particular Marxist interpretation of consumption that obscures rather than clarifies the processes involved. Castells was in fact concerned with the ways in which different social groups were *enabled* to consume goods and services, the means of provision rather than the form of consumption itself. The term remains, however, fraught with ambiguity and there is little point in agonizing over definitional problems. Moreover, this is ground that has been well covered already. Pinch (1989) concluded after detailed discussion that "the notion of collective consumption, together with the related concept of social reproduction, are both too complex to be encapsulated by any simple watertight definition" (p. 57). This conclusion does not lead Pinch to reject the use of the term (although consciously or otherwise its status is relegated to that of a "notion") but to argue for greater specification of particular forms of collective consumption in specific contexts and to adopt a dynamic approach rather than one that emphasizes static system needs. In practice it is not always evident what difference there is between collective consumption and more familiar terms such as collective state provision or simply state provision of welfare services. Indeed many authors use the terms interchangeably. The advantage in retaining the collective consumption perspective may be no more than its explicit link with theorization of the broader political economy of capitalist societies and the links between welfare provision and political struggle.

The concept of privatization is equally ambiguous, as a number of commentators have already observed (see, for example, Donnison, 1984;

Flynn, 1988; Lundqvist, 1989). Beyond the simple definition that it refers to the transfer of state activities to the private sector it requires more precise specification. As an economic concept, it emphasizes market provision, deregulation, denationalization, or contracting out. It may involve the privatization of the financing of a service that continues to be provided through the state (e.g., bank loans for students). It may refer to the privatization of the production of a service that continues to be state financed (e.g., private prisons). In the most extreme example it would involve a shift from a state regulated, publicly financed, publicly produced service to one that was fully deregulated and privately financed and produced. Such extreme transformations are actually difficult to imagine, although one could conceive, for example, of a major shift in the nature of law enforcement. Pirie (1988), a strong advocate of privatization, refers to it as a "complex and subtle process" and says that "each case is unique and requires a different remedy" (p. 11). He goes on to outline 21 methods of privatization ranging from "selling the whole by public share issue" to "the right to public substitution."

While the expansion of collective consumption conveys a general trend in the development of capitalist societies in the postwar period up until around the mid-1970s, privatization describes broad structural trends in a number of societies since then. It reflects the growing hegemony of New Right ideas in a period of fiscal austerity and economic restructuring. In that broad sense privatization and collective consumption are two sides of the same coin. The terms are rooted in rather different perspectives, however, and decollectivization or individualization might be a more appropriate polar opposite to collective consumption. Discussion of privatization also merges into broader sociological discussion of cultural privatism, home centeredness and self-provisioning (Franklin, 1989; Saunders & Williams, 1988). From this perspective it is not only capital that is released from the confines of an overextended state but the individual who escapes dependency and is enabled to realize his or her full potential as a free operator in an unfettered market. Postprivatization culture is portrayed as one in which workers drive home from their former state owned companies. They watch videos in their former state owned homes while selecting which school will benefit from their voucher. At the weekend they indulge in do-it-yourself activities and make home-made preserves. Hence privatization and collective consumption are not located at different points on one continuum but are multidimensional.

The social, cultural, and economic impact of particular cases of privatization will be differently perceived according to the nature and scale of the state activity involved. And this will vary cross-nationally. Critical differences will relate to whether the privatization is in the sphere of production (e.g., denationalization) or collective provision (e.g., health care); whether it affects directly households or the corporate sector; and whether it is majority or minority provision and the particular political history. For example, in the United Kingdom the social and cultural dimensions of deregulation of public transport or contracting out of hospital auxiliary services are less prominent than the extension of individual home ownership through selling state dwellings because of the scale of activity involved and the political history of housing tenure. What is important to emphasize is that the extent, nature, and distributional consequences of privatization have been determined by the previous pattern of state provision and intervention. The forms of state provision that developed in the postwar period contained within them different possibilities for their return to the market. Stated simply, one cannot deregulate what has not been regulated and the state cannot sell assets that it does not own.

EXPLANATION

Why has privatization happened? Responses to this question usually fall into two categories—those emphasizing system needs and those employing consumer preferences. The system needs explanation is rooted in the fiscal crises affecting local and national governments in the 1970s. The postwar period had seen the steady increase in public expenditure and public sector employment. Developing fiscal and profitability difficulties were partly attributed to the crowding out effects of a growing public sector—a view that the space for profitable private sector activity was being constrained by creeping state intervention. The solution was therefore to roll back the state sector and open up new avenues for entrepreneurial activity and profitable exploitation. Paradoxically, this view of the damaging effects of an expanding public sector on the health of a capitalist economy was entirely consistent with Left analyses. The difference was that Marxist critiques saw the problem as an inherent contradiction of capitalism. New Right critiques, however, analyzed the problem as one that capitalism could overcome through a reduction of state involvement in the economy and in social consumption. Another parallel was the similarity between the Left critiques of collective consumption in the 1960s and 1970 and the New Right critiques of the 1980s. From both perspectives state provision was portrayed as unresponsive, undemocratic, paternalistic, inegalitarian, for the benefit of producers rather than consumers, and about state ownership rather than democratic control (see, for example, Le Grande & Goodin, 1987; Offe, 1984). In other words, collective consumption was not just bad for the economy but bad for consumers.

Here the longstanding debates about collective consumption shade into a more recent revival of interest in consumer preferences and consumer culture. This developing literature links discussion of the politics of social provision and commodification to aspects of status display and emotional pleasures of consumption (see, for example, Featherstone, 1990). To some extent this has its roots in Castells's (1976) original formulations regarding the legitimate focus of urban sociology although the emphasis on consumption rather than production in the shaping of everyday experience has gained greater prominence recently in Saunders's call for an urban sociology that is a sociology of consumption, albeit with no spatial fix (Saunders, 1985, 1986). Gone, however, are references to urban (social) movements and the sphere of consumption as a site for political struggle. Instead, revealed preferences are elevated to a new importance. People do not struggle for better housing or better services. Those who can, exit, and opt for the privatized alternative. From this perspective there is no point in struggling for improved collective provision since state monopolies are inherently unresponsive to consumer needs. In a critique of Saunders, Warde (1990) suggests that he unquestioningly accepts that:

> expressed preference is the only yardstick of consumer behaviour. This effectively is to endorse the ideology of consumerism, consumerism being ideological to the extent that it misrepresents the social and collective determinants of taste, obscures supply side factors over marketing, conceals its own disappointing features etc. Saunders effectively endorses the ideology of consumerism. (p. 231)

Saunders' work meshes with more all encompassing claims that there has been a sea shift in philosophy through the failure of socialist ideas in practice. The literature emanating from institutions and individuals associated with various strands of New Right thinking has a strong evangelical tone (see, for example, Adam Smith Institute, 1986). In a fairly sober account of denationalization and deregulation in the United Kingdom, Veljanovski (1987) claims:

> Privatization is at the vanguard of a world-wide movement in thinking and politics about the legitimate role of the state in an industrial society in the 1980s. Socialism in whatever form has lost the battle of ideas and has been forsaken as a practical solution to the immediate industrial problems that most economies are now confronting. (p. 204)

This New Right evangelism can be found in various writings of authors such as Seldon (1977), Redwood (1988), and Pirie (1988) and in the various papers emanating from organizations such as the Adam Smith Institute and the Institute of Economic Affairs.

It is fair to say that many commentators who are unsympathetic to the New Right project readily acknowledge that state provision has entailed many of the failings identified by organizations such as the Adam Smith Institute. More recent writings that advocate pervasive privatization and deregulation as a response to those failings rarely acknowledge however that Marxist and neo-Marxist critiques of state welfare provision have already offered penetrating critiques of the provision. The Marxist literature of the 1960s and 1970s pointed to the dominance of producer and professional interests in the delivery of welfare services and the strong tendency for state provision to benefit the middle classes and the aristocracy of skilled manual workers. Far from eulogizing state provision the dominant tendency was to identify processes of segregation, ghettoization and class and racial divisions within the state sector itself. Those with least bargaining power in the labor market confronted state bureaucracies that allocated housing or other welfare benefits in a way that reflected and perpetuated divisions in the market. However, where Left critics part company with advocates of privatization is in their view that these defects were not the failings of state monopolies or socialism as such but rather the inadequacies of a system of state provision compromised by underfunding and the priorities of dominant interest groups. These compromises involved reductions in the quality of a service as it became more effectively targeted on the poorest groups and the often unanticipated and complex organizational dynamics of allocating in conditions of scarcity (see, for example, Henderson & Karn, 1987). Left critiques therefore pointed to the need for greater democratization, deprofessionalization, and decommodification.

Authors such as King (1989) and Esping Andersen (1985) have argued that the pressures for recommodification and the success of right wing policies have been greater where welfare provision is marginal rather than institutional in form. In other words, it is not the case that New Right policies have been most effective in countries where collective forms of provision have been most pervasive. The better established and more universal have been the social democratic forms of provision, the less open they have been to challenge. Thus it is argued that Britain and Denmark have seen more successful implementation of New Right policies than Norway and Sweden. Within this formulation the United Kingdom sits somewhat uneasily between the more deeply embedded welfare regimes such as Sweden and the more residual welfare forms found in the United States. In the United Kingdom substantial state owned assets such as housing were politically vulnerable and fiscally attractive. To regard the United Kingdom welfare state as marginal, however, is something of an oversimplification that underemphasizes the uneven impact of privatization within countries. In the United Kingdom, for example, the more universalistic forms of provision such as health care and state education have so far remained largely intact. It is those services that have an almost exclusively working-class clientele such as social security and state housing that have been cut or privatized most extensively.

It may be this combination of the political ascendancy of New Right ideas in a situation where there were substantial state assets to be privatized that explains the belief among U.K. commentators that it is the Thatcherite revolution that has led the way. While Milton Freidman has been one of the most influential ideologues and studies of deregulation such as that of Savas (1982) in the U. S. context have been important reference points for British analysts, privatization in practice has perhaps been most pervasive in the United Kingdom. Redwood (1988) suggests that "most would agree that the main influence and the dominant model for the rest of the world in the 1980s was the UK government's programme pieced together after the election of 1979" (p. 71). In a more colorful style a publication of the Adam Smith Institute claims that for overseas representatives of national governments

> there is almost a "privatization" tour as they meet people from the newly privatized industries, talk to the workforce, and meet with some of the customers. It is only a matter of time before some enterprising travel agent offers a package tour including visits to newly owned council houses, and inspections of garbage collection done by private companies. (1986, p. 11)

And in terms of the sale of state assets it is the United Kingdom that has developed the most extensive literature. Henig, Hamnett, and Feigenbaum (1988) comment that in France and Britain attention has been more focused on asset sales "while in the United States contracting out has been the most utilized vehicle" (p. 4). They comment further that:

> the learning process has not been unidirectional. The United States is seen by the other two nations as having provided a testing ground in contracting out of local services. And, both British and French officials seem to consider the US as something of a model when it comes to the establishment of a tradition of stock ownership among the large middle class. (p. 31)

The overall picture that emerges from the literature is of a combination of factors contributing to the erosion of collective consumption forms. And these factors have varied between countries and between sectors. In some cases the politics of state and market provision have had a long historical connection with established political parties and positions have been highly polarized. That embedded political polarization combined with fiscal constraint, a shift from quantitative to qualitative issues in service provision, the apparent failure of state provision to deliver and maintain a high-quality service, the rise of a new consumerism buttressed by surveys of consumer preferences, and a greater fragmentation of the working class have combined to greater or lesser degrees in different countries to produce a very different environment for state provision. Moreover as more households exit from the state sector it becomes increasingly the domain of the poorest sections of the working class and becomes caught in a spiral of decline and disinvestment. The old adage that services for poor people are invariably poor services is turned on its head and the service itself is blamed for the low social status of those who make use of it.

It is misleading, however, to exaggerate the scale of the changes that have occurred or their novelty. The rhetoric of the New Right should not be allowed to obscure the panoply of state subsidies that support and extend supposedly free market alternatives. Equally it is mistaken to overemphasize the extent to which collective consumption forms were ever free from the influences of market control, private finance, and private producers. The financing of collective consumption often involves borrowing on the private money market as market rates. And producer interests in, for example, pharmaceuticals in health care or private housebuilders in social housing sectors have been important influences in the shaping of welfare provision and the extent of decommodification.

PRIVATIZING STATE HOUSING

It is appropriate to explore these broader issues through a more detailed consideration of one area of state provision. It is also appropriate to focus on one particular form of privatization, asset sales in two rather different housing systems, namely the United Kingdom and the United States. Housing is a particularly good example for a number of reasons. Housing figures prominently in the literature on collective consumption. Historically it has been an important element in working-class struggles. Public housing has been the target of privatization policies across a wide range of nation states. Analyses and commentaries on existing or proposed programs of state housing privatization are available on a diversity of capitalist, state socialist and former state socialist, newly industrializing, and Third World societies. As well as the United Kingdom (e.g., Forrest & Murie, 1988) and the United States (e.g., Hayes, 1988; Legates, 1988; Stegman, 1990), these include Sweden (Danermark & Jacobsen, 1988; Lundqvist, 1988), West Germany (Krätke, 1989), the Netherlands (Boelhouwer & Van Weesep, 1988; Smeets & Martens, 1988), Hungary (Lowe & Tosics, 1988; Tosics, 1987), Yugoslavia (Caldarovic, 1988; Mandic, 1988), Israel (Werczberger & Reshef, 1990), Hong Kong (Castells, Goh, Kwok, & Kee, 1988) and China (Castells et al., 1988; Fong, 1988).

One reason for focusing on the United Kingdom is that the privatization of British state housing, which formed the vanguard of the broader privatization strategies of the Thatcher administrations, has had an important "demonstration" effect on other nation states. Stegman (1990) claims that the U.K. "Right to Buy" program in housing was "the ideological forerunner of the Reagan privatization scheme, and of HUD Secretary Jack Kempt's bill to stimulate the conversion of public housing to ownership by organized resident councils" (p. 10). And more than any other area it is supported by strong evidence on consumer preferences for a higher level of individual home ownership, a trend that has apparently international dimensions. It is a form of privatization that has most confused and compromised those who are generally critical of such policies in relation to other areas of state provision. And changes affecting the residential dwelling stock can have important fiscal impacts at both household and governmental levels as well as powerful spatial effects on residential differentiation and segregation.

One of the preconditions for an extensive program of privatization of state housing is to have state housing to sell. Here a stark contrast can be found between the United States where government has acted in a supplementary role to private enterprise and has favored subject rather than object subsidies (i.e., allowances and vouchers) and countries such as the United Kingdom where the main vehicle for the provision of working-class housing has been through direct provision by local authorities. In 1980 the United States had about 1.3 million public rental dwellings constituting 2% of the stock as a whole and 7% of all rental units. In the United Kingdom in the same year there were 6.5 million dwellings owned by local authorities, housing a third of all households and dwarfing the private rental sector. Public housing in the United States has always been strongly residual in character and the scope for asset sales has been extremely limited. Condominium conversion encouraged by tax regulations has been one of the principal ways of enabling sections of the working class to get into home ownership. This activity has been concentrated in the better quality buildings and neighbourhoods, however, and has had a limited impact on poorer households (Lundqvist, 1986). In Britain, however, the sale of state housing has been the principal route over the past decade for entry into home ownership for working-class households. By the end of the 1980s 1.5 million dwellings had been transferred from the public to the private sector, from collective to individual ownership. Not only does this represent a major transformation in the tenurial status of a large number of households, it represents a massive revalorization of the residential housing stock with associated fiscal gains to central government. Of all the privatization programs undertaken by the Thatcher administrations the sale of state owned dwellings to the tenants has been the most financially significant. While it is certainly the case that the program was initially ideologically driven, its fiscal importance has increased over time. Between 1979 and 1988 proceeds from all privatization programs (consisting mainly of denationalization of enterprises such as British Aerospace and British Petroleum) amounted to £32 billion (ca. $58 billion), of which 43% or some £14 billion came from the privatization of state housing. This revenue has been an important element in overall public expenditure planning by the Thatcher governments. Without the revenue from public housing privatization the scope for cuts in direct taxation would have been more limited. From 1988 to 1991 the official expectation is that the sale of state dwellings will deliver a further £9 billion. By any standards these are substantial amounts of money, especially when it is considered that public sector tenants in Britain now receive an average discount on market value of 50%. Public tenants living in apartments (rather than houses) can qualify for a maximum reduction on market value of 70%. One of the ironies of privatization is that it can be an expensive business for the state since it often involves the sale of assets at less than market value. While it is the working of the housing and labor markets that has excluded the households that are in the state sector, it is the state sector that is apparently to blame for their second class status. And the only way for households to move from state to the market sector is for there to be substantial interference with market pricing.

Some of these same issues have arisen in the housing privatization program in the United States although, as has already been stressed, the context is very different. The limited history of direct provision, and the limited amount of state housing assets available to sell off, has entailed a privatization program that has taken the form of fiscal retrenchment rather than "load shedding." Policies toward housing have been consistent with overall urban policy, a "new Federalism" that emphasized the rejuvenation of the national economy as the preeminent goal. Within that overall strategy, cities and their housing programs would either sink or swim. While various forms of privatization were prominent in the policies of the Reagan administrations, they were less widely implemented than planned or anticipated. For example, plans to sell vast tracts of state owned land were ultimately abandoned and plans to deregulate health care facilities were withdrawn. Barnekov, Boyle, and Rich (1989) suggest that privatization failures at the federal level resulted from an underestimation of the opposition to such policies.

As regards U.S. public housing, Legates (1988) points to five reasons why load shedding did not occur under the Reagan administration. First, there was little load to shed. Second, the physical design of the stock was inappropriate for individual or small-scale association ownership. Third, tenants were too poor. Fourth, a large number of tenants were too old or had major physical or mental disabilities. Finally, while there had been various demonstration projects, Congress regarded large-scale load shedding as neither feasible nor desirable.

A major difference between the U.S. and the U.K. situations, therefore, relates to the characteristics of the tenant populations. Those in favor of load shedding in the United Kingdom often referred to the number of tenants in the public sector who could afford market alternatives. There was a popular caricature of the affluent working-class household with three earners, an expensive car, and taking foreign holidays, that occupied a subsidized council house. This section of the working class, it was argued, should be encouraged or cajoled to leave the state sector. State housing should be for those who could not compete for housing in the private sector. The United Kingdom model should move nearer the U.S. model, a residual sector more effectively targeted on the poorest sections of the working class. In the United States by contrast there has been less scope to extend home ownership through the sale of state housing to tenants and no basis for a belief that the state sector contained a large number of affluent tenants. Public housing tenants in the United States

as a group have incomes that average less than one third the national median. . . . A full thirty-eighty percent of public housing residents are elderly. A majority of tenants have no working adults in the household which further reduces the potential for privatizing of America's public housing inventory. Of non-elderly public housing tenants in the U.S. . . . about half are single parent households with young children receiving public welfare benefits and only 42 percent contain one or more working adults. (Stegman, Rohe & Quercia, 1987, p. 10)

While much of the rhetoric in U.S. and U.K. housing policy has been very similar, those promoting privatization in the United States were faced with a generally impoverished population and a relatively unattractive dwelling stock. In contrast, much of the U. K. public housing stock was of relatively high quality, consisting of single family homes containing households that in many cases could have afforded to purchase in the private sector.

One question that arises in the latter situation is why households in the state sector who could have chosen to exit have not in fact done so. Consumer preference studies that show an overwhelming desire for home ownership suggest that more affluent public tenants would leave the state sector at the earliest opportunity. Without entering into an elaborate discussion of how such studies should be interpreted, it is evident that for individual households other considerations can override the desire to own. These relate to the desire to live in a particular locality, issues of kinship and social networks, and the superior use value that public housing can offer compared to what can be purchased elsewhere in the private sector.

Interpretation of the mass transfer of sitting tenants from the public to the private housing section in the United Kingdom through asset sales that suggest that it confirms dissatisfaction with collective consumption forms and their inherent defects, are quite misleading. Saunders and Harris (1987), for example, refer to the "high degree of dissatisfaction among tenants. . . . which lies at the root of their strong but often frustrated desire to 'exit' from the state sector" (p. 15). This view is, however, inconsistent with the majority of social surveys of tenants in the state sector in the United Kingdom, which show a relatively high degree of satisfaction. A survey in 1986 found that 77% of tenants were either very satisfied (32%) or fairly satisfied with their dwellings. And in 1984 a more detailed study of various aspects of public housing in England found that more than two thirds of tenants were satisfied with the service. None of the official surveys have found a majority of public tenants dissatisfied (for a more extensive review see Forrest & Murie, 1990). Moreover, there is a consistent body of evidence that has established that a substantial proportion of those who buy council houses as sitting tenants could have bought elsewhere in the city they lived in (for a review of this evidence see Forrest & Murie, 1988). In other words, if the experience of state housing and state landlordism are as negative as some critiques imply, why did substantial numbers of more affluent tenants not choose to exit from council housing and buy elsewhere? There is no paradox here. It is simply wrong to interpret the grievances of public tenants as evidence of endemic dissatisfaction with collective provision. Various studies present a consistent picture. Public tenants believe that the quality of the service could be improved. They do not believe that private landlordism would be any better. Indeed they believe, and often know, that it would be considerably worse. In general, they have a positive view of public housing but would prefer to be home owners. Their desire for house ownership does not, however, override other considerations. If tenants are offered the opportunity to purchase their dwelling at a heavily discounted price, many will choose to do so, but they will not move house in order to buy.

The fact that over a million state tenants in the United Kingdom have chosen to buy confirms their desire to own the dwelling they have made their home. Mass sales to dissatisfied tenants would be paradoxical. The sale of state dwellings in the United Kingdom has been legislatively possible almost since the inception of council housing. Local government could choose to operate discretionary policies and many did so. But until discounts were introduced virtually no sales occurred. The much smaller discounts of 20%-30% on market value that were on offer by some local authorities in the 1960s and 1970s did generate sales but nowhere near the scale of disposals achieved under the Thatcher administration. Under the Right to Buy, particularly for tenants occupying the most desirable dwellings, the high levels of discounts make it an offer that could not be refused.

At market value, therefore, the majority of tenants would not have chosen to buy and those on lower incomes could not have afforded to do so. It was not until large discounts were introduced in the 1980s that mass sales occurred. Similarly, it was estimated in the United States that at market value less than 1% of public units would be sold (Stegman et al., 1987, p. 9). The very market processes that are appealed to as superior in the justification of the policy are substantially undermined in its implementation. In the United Kingdom, discounts of up to 70% on market value are now offered as incentives and in order to make the dwellings affordable to poorer tenants. Such interference in market pricing would not be tolerated in other circumstances by those agencies and institutions involved in the housing market.

Comparisons of the United Kingdom and the United States illustrate, however, the way in which similar pro-privatization arguments are deployed against quite dissimilar regimes of collective consumption. The evident consequences of a housing policy that involves disinvestment in collective provision and a narrow concern with the further promotion of individual home ownership are, however, very similar. Legates (1988) offers an assessment of the consequences of recent U.S. policy that would be entirely appropriate to the United Kingdom. "Privatization's failure to improve the US housing supply or affordability situation has created high visibility problems at the local level. Most apparent are affordability problems for first time buyers, shortages and high cost of rental housing, and increasing homelessness for the very poor" (p. 193). While privatization appears the dominant policy trend across a wide spectrum of societies, it would be wrong to overemphasize ideological convergence, to neglect the widely varying housing circumstances, or indeed to propose a more general model of societal convergence in housing provision. There is a pervasive antistatism in housing policies, or at least a disillusionment with the kinds of mass housing solutions that are ideologically conflated with collective housing provision. Beyond that, however, the circumstances in which privatization policies in housing are being pursued or the directions in which they are taking specific welfare state regimes are widely varied.

OTHER FORMS OF HOUSING PRIVATIZATION

The previous section has focused particularly on the transfer of ownership of dwellings from the state as landlord to the individual consumer. There are other forms of privatization such as transfers of management functions, the shifting of rents toward market levels, changes in the nature of the producers of state housing, self-built, and so on. These developments are all evident in the various privatization strategies that have been pursued in different countries, as is the greater mobilization of family resources as state provision has been eroded (see, for example, Padovani, 1988; Sgritta, 1989). These other forms are, however, often less obvious transformations in collective consumption and often involve marginal changes in existing regimes rather than overt and explicit policy shifts with evident impacts on the living conditions of working-class households.

Asset sales along U.K. lines are occurring in countries such as Israel, France, and Austria and in Eastern Europe. But privatization in housing often refers to transfers of management functions from public to private agencies (Lundqvist, 1988). Where social housing has been provided through fixed term subsidy agreements, the expiry of such agreements can automatically trigger market reversion. This is most evidently the case in the Federal Republic of Germany where social housing objectives were generally achieved through granting public loans to institutions or individuals. Receipt of a state loan was accompanied by obligations regarding rent setting and building standards. After a certain period (generally 30 years) the public loan expires and the owner escapes the regulations. It has been estimated that on present policies half the social housing stock will have changed status by the early 1990s (see Krätke, 1989). A similar situation arises in the U.S. context with the progressive expiry of 40-year tax exempt bonds that have been the conventional means of providing low-income housing. Through an Annual Contributions Contract (ACC) between the public housing agency and the U.S. government, the U.S. Department of Housing and Urban Development pays the annual principal and interest of these bonds. When the contract for a particular housing project expires the decision regarding the future of those dwellings rests with the specific agency. This means that they could be privatized through various means without central government approval. Stegman et al. (1987) comment that "not many ACCs have expired at this point. As more do expire, however, the potential for privatization increases as does the potential threat of the loss of a community's permanent low-income housing resources" (p. 16).

Privatization may also refer to policies of building for sale with new units targeted at subsidized ownership rather than subsidized renting. This is, for example, the main element of Hong Kong's current privatization policies. And Fong (1988) refers to pilot schemes in China's special economic zones where small numbers of new units are sold to individuals. However, the principal means of housing privatization in China is through a raising of rents toward market levels. Castells et al. (1988) describe the process of housing privatization in Shenzen Special Economic Zone in terms that would seem radically interventionist and collectivist in a U.S. or U.K. context.

> Privatization of housing occurs both from the supplier and the consumer sides. The suppliers are organizationally and financially independent. The consumers—a combination of enterprises and workers—purchase housing services and pay full market price. The housing market is stratified and each submarket is monopolistic or oligopolistic in structure and in spatial distribution. Moreover, profits in housing are partially diverted to the state as revenue for local government investment and expenditure. . . . Enterprises, in order to improve workers' living standards and reduce their cost of living, either purchase or rent units from real estate companies and then provide housing for the managers and workers at extremely low rent or no rent at all. In Shekou, housing monopoly is total. Housing for workers and enterprises is supplied by only one real estate company, which provides all other construction in addition to housing. (p. 494)

The meaning of privatization in this context differs radically from the kinds of processes described in relation to the United States or United

Kingdom. In a similar vein, comparisons are often drawn between what is happening to housing systems in Eastern Europe and privatization in the United Kingdom, but this neglects major differences in the ownership and control of production, housing finance systems, and the continuing chronic housing shortages in, for example, Poland or Yugoslavia. Those differences illustrate a distinction between privatization and commodification in such contexts. Housing provision can be privatized in the sense that people have to find their own solutions through informal networks. In Yugoslavia, for example, there is discussion of asset sales of state owned dwellings to tenants but this may be more correctly seen in relation to a disorganized rather than overextended state. Caldarovic (1988) suggests that state intervention in the housing market in Yugoslavia has, in practice, been minimal and that living with parents or inheriting housing space as an important element has been the overall pattern of housing provision.

The importance of informal support of varying kinds (through the family or broader social networks) is evident in all countries and cautions against the evaluation of welfare regimes (and within that housing provision) along one dimension with the state and the market at either end. In the United Kingdom or the United States this is rather more obscured than in countries where the extended family and bonds of reciprocity remain important, such as in Greece, Italy, or France (see, for example, Tosi, 1988) or where self-build is prominent as in Hungary or Yugoslavia. In housing, therefore, the informal sector (and here I am including the voluntary sector) is an important element of the welfare package. Changes in patterns of state provision or increases or decreases in commodification interact with this third element in the equation. This can be interpreted and described in different ways depending upon the specific circumstances in which such processes are occurring. Where markets are underdeveloped it may be about the reduction of informal factors through greater marketization in production, distribution, and access. In the industrialized capitalist core countries the dominant discourse emphasizes greater self-reliance, self-help, self-build, and consumer empowerment. From some perspectives this is portrayed as shifting the burden onto the family and kinship networks as the state withdraws from collective provision. Alternatively, it is about debureaucratization and an escape from serfdom and dependency. Smeets and Martens (1988) refer to the ideological debate on privatization in the Netherlands and claim that there should be "more impulses for self-help. A 'careful society,' as stressed by the Christian democrats, should counterpose more and more the paralysing 'culture of dependency' " (p. 4). One reason for exploring the privatization of housing as an example of recent changes in collective state provision is that the promotion of individual home ownership has been argued to be a pivotal element in the liberation of working-class households from dependency and bureaucratic serfdom. This extends beyond the ideology that "dwellings are rented and homes are owned," or the acquisition of a potentially valuable asset by working-class households, to the function of the owned dwelling as an asset base for further involvement in privatized alternatives in, say, education or health care. High levels of home ownership can be mobilized to justify less generous pension schemes or disinvestment in state provided health care for an aging population. This scenario depends crucially on the continuing capital appreciation of residential dwellings. Recently, in the United Kingdom, this has become a more questionable assumption. Moreover, unlike other forms of privatization, state housing is absorbed rather differently into the market. The fact of previous state ownership clings to housing after it has been sold. Public housing privatization is therefore unlike selling shares in a state owned company. Eligibility is restricted to an almost exclusively working-class population, the sale is to the sole user of the service and, in most cases, the former state-owned elements remain identifiable from the rest (at least for the foreseeable future). In a depressed market the shares in home ownership are subject to even greater differentiation, and those that have been most recently decollectivized are likely to be the most stigmatized and to have low market values. In this sense the class basis of residential differentiation, which is obscured by the public/private distinction, reemerges as progressively greater proportions of urban residential structure become subject to market processes.

THE END OF COLLECTIVE CONSUMPTION?

This chapter began by suggesting that we were engaging with failed theory, false predictions about the structure of advanced capitalist societies, and the consequences of state disengagement from collective provision. The state had apparently withdrawn from various areas of provision without occasioning the negative impact on social cohesion that the theory seemed to predict. Arriving at firm conclusions is, however, complicated on a number of dimensions. First, the scale of the disengagement by the state may be exaggerated. The aspirations for pervasive privatization in welfare provision that have been particularly evident in the United Kingdom and United States have not yet been realized. Disengagement has been uneven. Second, the form rather than the scale of state support has changed with greater emphasis on the subsidy of individual rather than collective provision. In this area previous models of welfare support in advanced capitalist societies underestimated the possibilities of alternative welfare regimes. Third, the label of privatization used to describe a commonality of societal processes conceals widely divergent strategies, very different circumstances in which such strategies are being pursued, and a diversity of consequences and directions. Ideologues on both left and right have been guilty of generalizing across societies, understating diversity, and positing what Kemeny (1991) has described as "implicit theories of unilinearity and convergence" (p. 60).

It would be wrong, however, to suggest that little has changed as regards collective state provision, to neglect a dominant discourse of privatization, or to underplay the negative impacts on certain sections of the working class. There are new general conceptions to describe recent developments. Saunders (1986) has suggested that we are moving into a new epoch of individualized consumption and that the socialized, collective mode of consumption represented a transitional phase between a predominantly market mode and the rising privatized mode. He is careful to argue that there is nothing necessary or inevitable about such trends. Certain factors have combined (or at least so Saunders suggests in the U.K. context) to create the conditions for such developments. These include the crisis of socialized provision, the demand for privatized provision, and rising living standards (p. 315). This conception of recent developments does imply something of a deterministic relationship between rising living standards and decollectivization and overstates the extent to which there was a period of *predominantly* socialized provision. It also places great weight on consumer preferences as a major influence on modes of consumption, and understates the significance of supply side factors and underlying processes of commodification.

Some of the same themes emerge in the debates around post-Fordism and collective provision. This is an increasingly complex debate and only a brief reference is possible in this chapter. Stated simply, it is argued that "Fordism" is giving way to "post-Fordism," a regime of flexible as opposed to mass production. Associated with these changes is the dismantling of the previous collectivist Keynesian welfare statism (e.g., King, 1989). This view is summarized by Scott (1988) who states:

> The forward surge of the new flexible production sectors at the core of the new regime of accumulation has been further underpinned by major changes in the mode of social regulation. There has been a wholesale dismantling of the apparatus of Keynesian welfare-statism and deepening privatization of social life, a marked renewal of the forces of economic competition in industrial

production and labour markets and (in the United States) a sharp rise in governmental allocations for the purchase of military and space equipment. These changes are still in many ways in an experimental stage and no doubt their full extent and form remain yet to be determined. (p. 175)

Both the basic formulations and the claimed consequences of these developments have met with a number of challenges (see, for example, Costello, Michie, & Milne, 1989; Lovering, 1990). And Jessop (1990) has argued that rather than attempting to proceed from changes in the regime of accumulation to developments in the mode of regulation, it is the mode of regulation itself that should be the point of theoretical departure. Moreover, he suggests that in understanding the transition to post-Fordism we should be looking not only at state disengagement but also at the "rolling forward of a new type of state" (p. 33). It remains the case, nevertheless, that changes in the regime of accumulation have been rather more fully researched and articulated than the links between new spatial divisions of labor, changes in the production process, and new welfare state forms. The connections between the sphere of production and post-Fordist forms of welfare provision and support are often only loosely made (however, see Hoggett, 1990; Stoker, 1989). As Rustin (1989) has observed:

> The kind of Fordist fit between the mass production of goods and a similarly "massified" provision of public services is taken as the model of the relation we should be looking for between flexible, consumer-driven systems of welfare. . . . One implicit hypothesis is that consumers who are led to expect variety and choice in consumer markets will or should settle for nothing less in the welfare sector. (p. 59)

It is not clear how post-Fordist conceptions of changes in collective consumption can explain the apparent hegemonic discourse of privatization and state disengagement across such a diversity of nation states. Nor is it even the case that there has been a "wholesale" dismantling of collective welfare state forms in the United States or the United Kingdom. That may well be the direction of change but it is neither universal nor fully established. Nevertheless, the ideas surrounding the post-Fordist model of societal change seem to have opened up the most promising new avenues of exploration in relation to welfare provision.

For example, the conflation of "state" and "public" is now under scrutiny as are the basic organizational arrangements that have been traditionally associated with collective provision. This is the discourse of empowerment, decentralization, and democratic collectivism with the recognition that privatization and the commodification of urban services is only one of many directions. Hoggett (1990) argues that a quite different future is possible that is equally critical of the power of professional interests, waste, inefficiency, and the lack of responsiveness and accountability in state provision. Technological developments have created new forms of, and new possibilities for, organizational control and structure but the shift toward the externalization of production in urban service provision is not inherently Thatcherite.

> The New Right has sought to harness this through a strategy based upon the hegemony of the commercial contract positioned within a set of market rules whose intention is to cost control, the bifurcation of the public service labour market and the dissolution of localised forms of democratic accountability. In contrast, the principle of external decentralisation could be developed to construct a welfare system which sought to achieve a real form of pluralistic and democratic collectivism, one which genuinely struggled to achieve a balance between state institutions on the one hand and self-managed communities on the other. (Haggett, 1990, p. 50)

We should not be mesmerized, however, by the rhetoric of decentralization, flexibilization, self-management, and democratic accountability and neglect the continuing concentration and centralization of capital and the way in which collective state provision has always been heavily compromised and circumscribed by the logic of the market. To return to the housing sector, in western capitalist societies market reversion was often built into social housing forms and this cautions against seeing the current phase of privatization as a sharp rupture rather than as an intensification of underlying trends. Also, housing directly provided and owned by the state (as in the United Kingdom) was typically built by the private construction industry and financed by private capital and the shifts in its form owed as much if not more to the vicissitudes of capitalism and the relative bargaining power of sections of the working class as to the organizational dynamics of state bureaucracies. Moreover, we should not assume that the recasting of sections of the working class as active consumers rather than as passive state recipients will mean that those who cannot compete effectively in the new privatized cities will now get what they want rather than what they are given. The transformations of collective provision may have been partial and uneven but there have been real and major effects, particularly in social housing provision that have had (or will have) widespread impacts on those sections of the working class that lack strong political and economic bargaining power. It is no accident that it is groups such as the working-class elderly, single-parent families, ethnic minorities, the unemployed, and the insecurely employed that have felt the greatest negative impacts of privatization and disinvestment. In parallel with an international discourse of privatization and commodification is one of the ghettoization, polarization, segregation, and exclusion. There may be a diversity of processes at work but there is a commonality of consequences. As residential space is privatized and progressively commodified, as the dynamics of privatization gather pace, cities are experiencing increasing squalor and homelessness, displacement, and decay. It may therefore be premature to pass firm judgments on the theoretical models discussed earlier. The consequences of privatization and disinvestment are only now beginning to become apparent. It is possible that struggles against those processes will emerge and intensify and that we will enter a phase where the emphasis shifts toward new forms of collective provision and away from individualistic, market-oriented policies. Whether or not that occurs, theory will need to transcend unilinear models of "market" and "state," the untenable distinctions between "public" and "private," and be more sensitive to diversity in welfare regimes.

REFERENCES

Adam Smith Institute. (1986). *Privatization worldwide*. London: Adam Smith Institute.

Barnekov, T., Boyle, R., & Rich, D. (1989). *Privatism and urban policy in Britain and the United States*. Oxford, UK: Oxford University Press.

Boelhouwer, P., & Van Weesep, J. (1988). The sale of public housing and the social structure of neighbourhoods. *Built Environment, 14*(3/4), 145-154.

Caldarovic, O. (1988, September). *Concepts of housing and forms of privatization in Yugoslavia*. Paper presented at the Research Conference "Housing Between State and Market," Dubrovnik.

Castells, M. (1976). *The urban question*. London: Edward Arnold.

Castells, M., Goh, L., & Kwok, R., with Kee, T. L. (1988). Economic development and housing policy in the Asian Pacific Rim: A comparative study of Hong Kong, Singapore and Shenzhen Special Economic Zone. *Monograph 37*. Berkeley: University of California, Institute of Urban and Regional Development.

Costello, N., Michie, J., & Milne, S. (1989). *Beyond the casino economy*. London: Verso.

Danermark, B., & Jacobsen, T. (1988, September). *Privatisation and residential segregation*. Paper presented at the Research Conference. "Housing Between State and Market," Dubrovnik.

Donnison, D. (1984). The progressive potential of privatization. In J. Le Grand & R. Robinson (Eds.), *Privatization and the welfare state*. London: Allen & Unwin.

Esping Andersen, G. (1985). Power and distributional regimes. *Politics and Society, 4,* 223-256.

Featherstone, M. (1990). Perspectives on consumer culture. *Sociology, 24*(1), 5-22.

Flynn, R. (1988). Political acquiescence, privatization and residualisation in British housing policy. *Journal of Social Policy 17*(3), 289-312.

Fong, F. N. (1988). *Housing reforms in China: Privatization of public housing in a socialist economy.* Hong Kong: University of Hong Kong, Centre of Urban Studies and Urban Planning.

Forrest, R., & Murie, A. (1988). *Selling the welfare state: The privatization of public housing.* London: Routledge & Kegan Paul.

Forrest, R., & Murie, A. (1990). A dissatisfied state? Consumer preferences and council housing in Britain. *Urban Studies, 27*(5), 617-635.

Franklin, A. (1989). Working class privatism: An historical study of Bedminster, Bristol. *Environment and Planning D: Society and Space, 7,* 93-113.

Harloe, M., & Paris, C. (1984). The decollectivization of consumption: Housing and local government finance in England and Wales 1979-1981. In I. Szelenyi (Ed.), *Cities in recession.* London: Sage.

Harrison, M. L. (1986). Consumption and urban theory: An alternative approach based on the social division of welfare. *International Journal of Urban and Regional Research, 10,* 232-242.

Hays, R. A. (1988). Housing and the future of the American welfare state. *Built Environment, 14*(3/4) 177-189.

Henderson, J., & Karn, V. (1987). *Race, class and state housing.* Aldershot, UK: Gower.

Henig, J. R., Hamnett, C., & Feigenbaum, H. B. (1988). The politics of privatization: A comparative perspective. *Governance, 1* (4), 442-468.

Hoggett, P. (1990). Modernisation, political strategy and the welfare state: An organisational perspective. *Studies in decentralisation and quasi-markets, 2.* Bristol, UK: University of Bristol, School for Advanced Urban Studies.

Jessop, B. (1990, March). *Fordism and post Fordism* [A revised version of a paper presented to the Conference on Pathways to Industrialization and Regional Development, Lake Arrowhead, CA].

Kemeny, J. (1991). *Housing and social theory.* London: Routledge, & Kegan Paul.

King, D. (1989). Economic crisis and welfare state recommodification: A comparative analysis of the United States and Britain. In M. Gottdiener, & N. Komninos (Eds.), *Capitalist development and crisis theory: Accumulation, regulation and spatial restructuring* (pp. 237-260). London: Macmillan.

Krätke, S. (1989). The future of social housing—problems and prospects of "social ownership": The case of West Germany. *International Journal of Urban and Regional Research, 13*(2), 282-303.

Le Grand, J., & Goodin, R. (1987). *Not only the poor, Middle classes and the welfare state.* London: Allen & Unwin.

Legates, R. (1988). The local government backlash against federal housing privatization in the United States. *Built Environment, 14*(3/4), 190-200.

Lovering, J. (1990, May). A perfunctory sort of post Fordism: Economic restructuring and labour market segmentation in Britain in the 1980s. *Work, Employment and Society.* [Special Issue].

Lowe, S., & Tosics, I. (1988). The social use of market processes in British and Hungarian housing policies. *Housing Studies, 3*(3), 159-171.

Lundqvist, L. (1986). *Housing policy and equality.* London: Croom Helm.

Lundqvist, L. (1988, September). *Corporatist implementation and legitimacy: The case of privatization in Swedish public housing.* Paper presented to the Research Conference "Housing Between State and Market," Dubrovnik.

Lundqvist, L. (1989). Explaining privatization: Notes towards a predictive theory. *Scandinavian Political Studies, 12*(2), 129-145.

Mandic, S. (1988, September). *Forms of privatization in Yugoslavia.* Paper presented to Research Conference "Housing between State and Market," Dubrovnik.

Offe, C. (1984). *Contradictions of the welfare state*. London: Hutchinson.

Padovani, L. (1988, September). *Housing provision in Italy: The family as emerging promoter. Difficult relationships with public policies.* Draft paper presented to the Research Conference "Housing between State and Market," Dubrovnik.

Pahl, R. E. (1977). Collective consumption and the state in capitalist and state socialist society. In R. Scase (Ed.), *Industrial society: Class cleavage and control.* London: Tavistock.

Pinch, S. (1989). Collective consumption. In J. Wolch & M. Dear (Eds.), *The Power of Geography: How territory shapes social life.* London: Unwin Hyman.

Pirie, M. (1988). *Privatization: Theory, practice and choice.* Aldershot, UK: Wildwood House.

Redwood, J. (1988). *Popular capitalism.* London: Routledge & Kegan Paul.

Rustin, M. (1989). The shape of new times. *New Left Review, 175,* 54-77.

Saunders, P. (1985). Space, the city and urban sociology. In D. Gregory & J. Urry (Eds.), *Social relations and spatial structures.* New York: St. Martin's.

Saunders, P. (1986). *Social theory and the urban question.* London: Hutchinson.

Saunders, P., & Harris, C. (1990). Privatization and the consumer. *Sociology, 24*(1), 57-75.

Saunders, P., & Harris, C. (1987, September). *Biting the nipple? Consumers preferences and state welfare.* Paper presented at Sixth Urban Change and Conflict Conference, University of Kent, England.

Saunders, P., & Williams, P. (1988). The constitution of the home: Towards a research agenda. *Housing Studies, 3*(2), 81-93.

Savas, E. S. (1982). *Privatizing the public sector.* Chatham, NJ: Chatham House.

Scott, A. J. (1988). Flexible production systems and regional development: The rise of new industrial spaces in North America and western Europe. *International Journal of Urban and Regional Research, 12*(2), 171-186.

Seldon, A. (1977). *Charge.* London: Temple Smith.

Sgritta, G. (1989). Towards a new paradigm: Family in the welfare state crisis. In K. Boh et al. (Eds.), *Changing patterns of European family life.* London: Routledge & Kegan Paul.

Smeets, J., & Martens, M. (1988, September). *Forms of privatization and some effects on the Dutch system of housing provision.* Paper presented to the Research Conference "Housing between State and Market," Dubrovnik.

Stegman, M. (1990). *Privatizing public housing: Getting the government out of the way or out of the business?* Report prepared for the Twentieth Century Fund.

Stegman, M., Rohe, W., & Quercia, R. (1987). *US experience with the privatization of public housing.* Paper prepared under contract to PADCO Services Inc, Washington, DC.

Stoker, G. (1989, September). *Restructuring local government for a post-Fordist society?* Paper presented at the Seventh Urban Chance and Conflict Conference, University of Bristol.

Szelenyi, I. (1981). Structural changes of and alternatives to capitalist development in the contemporary urban and regional system. *International Journal of Urban and Regional Research, 5,* 1-14.

Titmuss, R. (1976). *Essays on the welfare state.* London: Allen & Unwin.

Tosi, A. (1988, June). *Informal housing practices in industrialised countries: The role of extended family networks.* Paper prepared for the International Conference on Housing Policy and Urban Innovation, Amsterdam.

Tosics, I. (1987). Privatization in housing policy: The case of the western countries and that of Hungary. *International Journal of Urban and Regional Research, 11*(1), 61-78.

Veljanovski, C. (1987). *Selling the state: Privatization in Britain.* London: Weidenfeld & Nicolson.

Warde, A. (1990). Production, consumption and social change: Reservations regarding Peter Saunders' sociology of consumption. *International Journal of Urban and Regional Research, 14*(2), 228-248.

Werczberger, E., & Reshef, N. (1990, July). *The privatization of public housing in Israel.* Paper presented to the International Housing Research Conference, Paris.

9

Partnership and the Pursuit of the Private City

GREGORY D. SQUIRES

> . . . two hundred and sixty-eight years of laissez-faire economics had left the city in a hell of a mess.
>
> Joseph S. Clark, Jr.,
> Mayor of Philadelphia 1952-1956
> (Warner, 1987, p. xi)

Public-private partnerships have become the rallying cry for economic development professionals throughout the United States (Davis, 1986; Porter, 1989). As federal revenues for economic development, social service, and other urban programs diminish such partnerships are increasingly looked to as the key for urban revitalization (G. Peterson & Lewis, 1986). These partnerships take many forms. Formal organizations of executives from leading businesses have been established that work directly with public officials. In some cases public officials as well as representatives from various community organizations are also members. Some partnerships have persisted for decades working on an array of issues while others are ad hoc arrangements that focus on a particular time-limited project. Direct subsidies from public agencies to private firms have been described as public-private partnerships. If economic development has emerged as a major function of local government, public-private partnerships are increasingly viewed as the critical tool.

The concept of partnership is widely perceived to be an innovative approach that is timely in an age of austerity. In fact, "public-private partnership" is little more than a new label for a long-standing relationship between the public and private sectors. Growth has been the constant, central objective of that relationship, though in recent years subsidization of dramatic economic restructuring has become a complementary concern. While that relationship has evolved throughout U.S. history, it has long been shaped by an ideology of privatism that has dominated urban redevelopment from colonial America through the so-called postindustrial era (Barnekov, Boyle, & Rich, 1989; Krumholz, 1984; Levine, 1989; Warner, 1987). The central tenet of privatism is the belief in the supremacy of the private sector and market forces in nurturing development, with the public sector as a junior partner whose principal obligation is to facilitate private capital accumulation. Individual material acquisitiveness is explicitly avowed, but that selfishness is justified by the public benefits that are assumed to flow from the dynamics of such relations.

One need look no further than the roadways, canals, and railroads of the eighteenth and nineteenth centuries to see early concrete manifestations of large-scale public subsidization of private economic activity and the hierarchical relationship between the public and private sectors (Krumholz, 1984; Langton, 1983). These relationships crystallized in the urban renewal days of the 1950s and 1960s and the widely celebrated partnerships of the 1980s. Structural changes in the political economy of cities, regions, and nations altered the configuration of specific public-private partnerships, but not the fundamental relationship between the public and private sectors. These structural changes have, however, influenced the spatial development of cities and exacerbated the social problems of urban America.

The continuity reflected by public-private partnerships, despite some new formulations in recent years, is revealed by the persistence in the corporate sector's efforts to utilize government to protect private wealth, and primarily on its terms. Demands on the state to subsidize a painful restructuring process have placed added strains on public-private relations. The glue that holds these efforts together, despite these tensions, is the commitment to privatism.

Focusing on the postwar years, this chapter examines the ideology of privatism, its influence on the evolution of public-private partnerships, and their combined effects on the structural, spatial, and social development of cities in the United States, and the lives of people residing in the nation's urban neighborhoods. Perhaps the most striking feature of the evolution of American cities, to be explored in the following pages, is the uneven nature of urban development. To many, such uneven development simply reflects the "creative destruction" that Schumpeter (1942) asserted was essential for further economic progress in a capitalist economy. To others, however, the unevenness generated by unrestrained market-based private capital accumulation constitutes the core of the nation's urban problems. After reviewing the theoretical debates over privatism, various contours of uneven development and the role of partnerships in particular and privatism generally in nurturing such inequalities are examined. Industrial restructuring and uneven spatial development of urban America, along with the many social costs associated with such development, are delineated, with a particular focus on the changing dynamics of racial inequality. Drawing from data pertaining to national trends in urban development as well as developments within specific cities, the mutually reinforcing effects of race and class are explored. This chapter concludes with a discussion of recent challenges to the ideology of privatism. City dwellers, many community organizations, and a significant number of public officials have begun to develop specific policy alternatives, including more inclusive partnerships, in hopes of achieving something better than the "mess of laissez-faire."

PRIVATISM

The American tradition of privatism was firmly established by the time of the Revolution in the 1700s. According to this tradition individual and community happiness are to be achieved through the search for personal wealth. Individual loyalties are to the family first, and the primary obligation of political authorities is to "keep the peace among individual money-makers" (Warner, 1987, p. 4). Always implicit, and frequently explicit, from colonial days to the present has been the primacy of private action and actors.

Consistent with free market, neoclassical economic theory generally, theory and policy in economic development and urban redevelopment circles have focused on private investors and markets as the appropriate dominating forces. Private economic actors are credited with being the most productive, innovative, and effective. Presumably neutral and impersonal market signals are deemed the most efficient and therefore appropriate measures for determining the allocation of economic resources. Given Adam Smith's invisible hand, the greatest good for the greatest numbers is achieved by nurturing the pursuit of private wealth.

Public policy, from this perspective, should serve private interests. Government has an important role, but one that should focus on the facilitation of private capital accumulation via the free market. (Privatism should not be confused with privatization. The former refers to a broader ideological view of the world generally and relationships between the public and private sectors in particular. The latter constitutes a specific policy of transferring ownership of particular industries or services from government agencies to private entrepreneurs.) While urban policy must acknowledge the well-known problems of big cities, it can do so best by encouraging private economic growth. A critical assumption is that the city constitutes a unitary interest and all citizens benefit from policies that enhance aggregate private economic growth (P. Peterson, 1981). Explicit distributive or allocational choices are to be avoided whenever possible, with the market determining where resources are to be directed. Public policy should augment but not supplant market forces (Barnekov et al., 1989; Levine, 1989).

The ideology of privatism has been tested in recent years by regional shifts in investment and globalization of the economy in general that have devastated entire communities (Bluestone & Harrison, 1982; Eisinger, 1988). Advocates of privatism attribute such developments primarily to technological innovation and growing international competition. They claim the appropriate response is to accommodate changes in the national and international economy. Given that redevelopment is presumed to be principally a technical rather than political process, cities must work more closely with private industry to facilitate such restructuring in order to establish more effectively their comparative advantages and market themselves in an increasingly competitive economic climate. Such partnerships, it is assumed, will bring society's best and brightest resources (which reside in the private sector) to bear on its most severe public problems.

Where such efforts cannot succeed cities must adjust, which in some cases means to downsize, just like their counterparts in the private sector. So-called pro-people rather than pro-place policies are offered to help individuals accommodate such changes. These adjustments may well mean moving from one city and region to another. Policies that might intervene in private investment decision making or challenge market forces for the betterment of existing communities are explicitly rejected (Kasarda, 1988; McKenzie, 1979; President's Commission for a National Agenda for the Eighties, 1980).

Concretely, the policies of privatism consist of financial incentives to private economic actors that are intended to reduce factor costs of production and encourage private capital accumulation, thus stimulating investment, ultimately serving both private and public interests. The search for new manufacturing sites, retooling of obsolete facilities, and restructuring from manufacturing to services have all been facilitated by such subsidization. During the postwar years cities have been dramatically affected by the focus on downtown development that has generally taken the form of office towers, luxury hotels, convention centers, recreational facilities, and other paeans to the postindustrial society. Real estate investment itself is frequently viewed as part of the antidote to deindustrialization. All of this is justified, however, by the assumption that a revitalized economy generally and a reinvigorated downtown in particular will lead to regeneration throughout the city. As more jobs are created and space is more intensively utilized, more money is earned and spent by local residents, new property and income tax dollars bolster local treasuries, and new wealth trickles down throughout the metropolitan area. Among the specific policy tools are tax abatements, low-interest loans, land cost writedowns, tax increment finance districts (TIFS), enterprise zones, urban development action grants (UDAGs), industrial revenue bonds (IRBs), redevelopment authorities, eminent domain, and other public-private activities through which private investment is publicly subsidized. The object of such incentives, again, is the enhancement of aggregate private economic growth by which it is assumed the public needs of the city can be most effectively and efficiently met.

Privatism has been a powerful ideological force in all areas of American life. That it has dominated urban policy should come as no surprise. But the pursuit of the private city has had its costs. And the advocates of privatism have had their critics.

RESPONSES TO PRIVATISM

The most fundamental intellectual and political challenges to privatism are directed to its central assumptions regarding the neutrality and imperson-ality of the market. Rather than viewing the market as a mechanism through which random decisions made by many individual willing buyers and sellers yields the most efficient production and distribution of resources for cities and society generally, it is argued that the market is an arena of conflict. Logan and Molotch (1987) observe that markets themselves are cultural artifacts bound up with human interests. Markets are structured by, and reflect differences in, wealth and power. They reinforce prevailing unequal social relations and dominant values, including a commitment to privatism. Markets are not simply neutral arbiters maximizing efficiency in production and distribution. They are social institutions firmly embedded in the broader culture of American society.

A related critique of privatism is the argument that a city does not constitute a unitary interest that can best be advanced through aggregate private economic growth, but rather a series of unequal and conflictual interests, some of which are advanced through a political process. As Stone (1987) has argued, local economic development policy represents the con-scious decisions made by individuals with highly unequal power in a community in efforts by competing groups to further their own interests. Assumptions of a unitary interest or the benevolence of market-based allocation mystifies important decisions made at the local level that clearly favor some interests at the expense of others. Development, therefore, is not a technical problem but rather a political process. As Stone concludes, "urban politics still matters" (1987, p. 4).

While economic development and urban redevelopment are political matters, one consequence of the pursuit of the private city has been a reduction in the public debate over development policy and the accountability of public officials and other actors for the consequences of their activities. Quasi-public redevelopment authorities have provided selected private investors with responsibilities traditionally vested in the public sector. Hidden incentives have been provided through such off-budget subsidies as industrial revenue bonds and bailouts for large but failing firms. Eminent domain rights have been granted to and exercised for private interests where public interests are most vaguely identified (Barnekov et al., 1989). The beneficiaries of these policies include real estate developers, commercial business interests, manufacturers, and others who view the city primarily in terms of the exchange value of its land at the expense of the majority for whom the city offers important use values as a place to live, work, and play (Logan & Molotch, 1987). But it is not just the immediate beneficiaries who share this view of local governance. As Gottdiener concluded, "The reduction of the urban vision to instrumental capital growth, it seems, gains hegemony everywhere" (Gottdiener, 1986, p. 287).

Declining accountability may be a factor contributing to a more concrete challenge to privatism. Simply put, it has not worked. That is, the array of subsidies and related supply-side incentives have not created the anticipated number of jobs or jobs for the intended recipients, tax revenues have not been stabilized as initially expected, and the urban renaissance remains, at best, a hope for the future (Barnekov et al., 1989; Center for Community Change, 1989; Levine, 1987). While not always ineffective, such incentives are not primary determinants of private investment decisions. And they often embody unintended costs resulting in minus-sum situations as public subsidies outrun subsequent public benefits (Eisinger, 1988, pp. 200-224). One reason for the disappointing results is that with the proliferation of incentives, the competitive advantage provided by any particular set of subsidies is quickly lost when other communities match them. The number of state location incentive programs alone increased from 840 in 1966 to 1,633 in 1985 (Eisinger, 1988, p. 19). Indeed, many states and municipalities feel obligated to offer additional incentives of acknowledged questionable value simply to keep up with their neighbors and provide symbolic assurance that they offer a good business climate. As Detroit Mayor Coleman Young observed:

> Those are the rules and I'm going by the goddamn rules. This suicidal outthrust competition among the states has got to stop but until it does, I mean to compete. It's too bad we have a system where dog eats dog and the devil takes the hindmost. But I'm tired of taking the hindmost. (Greider, 1978)

Ironically, one of the costs is the reduced ability of local municipalities to provide the public services that are far more critical in assuring a favorable climate for the operation of successful businesses. Tax dollars that are utilized as subsidies for private development are dollars that are not available for vital public services. In Detroit, for example, the quality of public education has declined precipitously in recent years, undercutting the ability of that city's youth to compete for jobs and the city's ability to attract employers (Thomas, 1989).

Another factor contributing to the disappointing results strikes at the heart of the ideology of privatism. As Bluestone and Harrison (1982) concluded in discussing such approaches to reindustrialization, "all share a studied unwillingness to question the extent to which conventional private ownership of industry and the more-or-less unbridled pursuit of private profit might be the causes of the problem" (1982, p. 230).

If privatism has not generated the anticipated positive outcomes, economic restructuring associated with privatism has generated a host of social costs that are either ignored or accepted by its proponents as an inevitable price to be paid for progress. Job loss and declining family income resulting from a plant closing are just the most obvious direct costs. There are also "multiplier effects." Economic stress within the family often leads to family conflicts, including physical abuse, frequently culminating in divorce. Increasing physical and medical health problems, including growing suicide rates, have been clearly connected to sudden job loss. The economic stability of entire communities and essential public services have been crippled (Bluestone & Harrison, 1982). Even the winners of the competition have suffered severe social costs. Sudden growth has generated unmanageable traffic congestion and skyrocketing housing costs often forcing families out of their homes and business to pay higher salaries for competent employees (Dreier, Schwartz, & Greiner 1988). Gentrification moves many poor people around but does little to reduce poverty. Even in Houston, the "free enterprise city," sudden private economic growth has generated serious problems in sewage and garbage disposal, flooding, air and water pollution, congestion, and related problems (Feagin, 1988). Perhaps the most destructive aspect of this "creative" process is the uneven nature of the spatial development of cities and the growing inequality associated with race and class (Bluestone & Harrison, 1988). Privatism and the economic restructuring that it has nurtured have created costs that are quite real, but not inevitable. As the critics of privatism note, they reflect political conflicts and political decisions (discussed in the following section), not natural outcomes of ultimately beneficial market forces.

The pursuit of the private city appears to have produced many ironies. Given the array of incentives, those firms intending to expand or relocate anyway often shop around for the best deal they can get. Consequently, local programs designed to leverage private investment are turned on their head. That is, the private firms leverage public funds for their own development purposes; and they can punish local governments that are not forthcoming with generous subsidies. A logical consequence of these developments is that private economic growth has become its own justification. As William E. Connolly observed:

> at every turn barriers to growth become occasions to tighten social control, to build new hedges around citizen rights, to insulate bureaucracies from popular pressures while opening them to corporate influence, to rationalize work processes, to impose austerity on vulnerable constituencies, to delay programs for environmental safety, to legitimize military adventures abroad. Growth, previously seen as the means to realization of the good life, has become a system imperative to which elements of the good life are sacrificed. (1983, pp. 23-24)

But perhaps these outcomes are not ironic. In fact, they may well be the intended results. As Barnekov et al. concluded in evaluating privatism in the 1980s, "The overriding purpose of the 'new privatism' was not the regeneration of cities but rather the adaptation of the urban landscape to the spatial requirements of a post-industrial economy" (1989, p. 12). That adaptation has been the central objective of public-private partnerships in the "postindustrial" age of urban America.

The postwar debate over privatism, like debates over redevelopment in general, have taken place within the context of dramatic structural changes in the political economy of American cities. The spatial development of urban America has clearly been influenced by these changes. In turn, the structural and spatial developments of cities have given rise to a host of social problems with which policymakers continue to wrestle. These struggles have included efforts to challenge the ideology and politics of privatism; challenges that have met with some success, including capturing the mayor's office in a few major cities. These efforts are discussed in the concluding section. The following section examines the evolving dynamics of privatism and partnerships for urban America during the postwar years, developments that have prepared the ground for challenges to privatism in recent years.

STRUCTURAL, SPATIAL, AND SOCIAL DEVELOPMENT

URBAN RENEWAL AND THE PROSPEROUS POST-WAR YEARS

The United States emerged from World War II as a growing and internationally dominant economic power. Given its privileged structural position at that time, the end of ideology was declared and optimism for future growth and prosperity was widespread (Bell, 1960).

Yet blighted conditions within the nation's central cities posed problems for residents trapped in poverty and for local businesses threatened by conditions within and immediately surrounding the downtown business center. Recognizing the "higher uses" (i.e., more profitable for developers and related businesses) for which such land could be utilized, a policy of urban renewal evolved that brought together local business and government entities in working partnerships with the support of the federal government. At the same time, federal housing policy and highway construction stimulated homeownership and opened up the suburbs, while reinforcing the racial exclusivity of neighborhoods.

As Mollenkopf (1983) has observed, urban renewal and related federal programs reflected a political coalition of disparate groups. Local entrepreneurial Democratic politicians, along with their counterparts at the federal level, created large-scale downtown construction projects that benefited key local contractors and unions, machine politicians and reformers, and white ethnic groups along with at least some racial minorities. These emerging political alliances were clearly, though not always explicitly, committed to economic growth (particularly downtown) with the private sector as the primary engine for, and beneficiary of, that development (Mollenkopf, 1983).

Although urban renewal was launched and initially justified as an effort to improve the housing conditions of low-income urban residents, it quickly became a massive public subsidy for private business development, particularly downtown commercial real estate interests (Barnekov et al., 1989, pp. 39-48; Hays, 1985, pp. 173-191). Shopping malls, office buildings, and convention centers rather than housing became the focus of urban renewal programs. Following the lead of the Allegheny Conference on Community Development formed in Pittsburgh in 1943, coalitions of local business leaders were organized in most large cities to encourage public subsidization of downtown development. Examples include the Greater Milwaukee Committee, Central Atlanta Progress, Inc., Greater Philadelphia Movement, Cleveland Development Foundation, Detroit Renaissance, the Vault (Boston), the Blyth-Zellerback Committee (San Francisco), Greater Baltimore Committee, and Chicago Central Area Committee. Using their powers of eminent domain, city officials generally would assemble land parcels and provide land cost writedowns for private developers. In the process local business associations frequently operated as private governments as they designed and implemented plans that had dramatic public consequences but did so with little public accountability.

If such developments were justified rhetorically as meeting important public needs, indeed urban renewal took sides. Not all sides were represented in the planning process and the impact of urban renewal reflected such unequal participation (Friedman, 1968). Some people were forcefully relocated so that others could benefit. According to one estimate, by 1967 urban renewal had destroyed 404,000 housing units, most of which had been occupied by low-income tenants, while just 41,580 replacement units for low- and moderate-income families were built (Friedland, 1983, p. 85). As Chester Hartman concluded, "the aggregate benefits are private benefits that accrue to a small, select segment of the city's elite 'public,' while the costs fall on those least able to bear them" (Hartman, 1974, p. 183).

At the same time that the public sector was subsidizing downtown commercial development, it was also subsidizing homeownership and highway construction programs to stimulate suburban development. Through Federal Housing Administration (FHA), Veterans Administration (VA) and related federally subsidized and insured mortgage programs launched around the war years, long-term mortgages requiring relatively low down payments made home ownership possible for many families who previously could not afford to buy. With the federal insurance, lenders were far more willing to make such loans (Hays, 1985; Jackson, 1985). (An equally if not more compelling factor leading to the creation of these programs was the financial assistance they provided to real estate agents, contractors, financial institutions, and other housing related industries [Hays, 1985]). Since half the FHA and VA loans made during the 1950s and 1960s financed suburban housing, the federal government began, perhaps unwittingly, to subsidize the exodus from central cities to suburban rings that characterized metropolitan development during these decades (Hays, 1985, p. 215). The Interstate Highway Act of 1956, launching construction of the nation's high-speed roadway system, further subsidized and encouraged that exodus.

A significant feature of these developments was the racial exclusivity that was solidified in part because the federal government encouraged it. Through the 1940s the FHA's underwriting manuals warned of "inharmonious racial or nationality groups" and maintained that "if a neighborhood is to retain stability, it is necessary that properties shall continue to be occupied by the same social and racial classes" (Jackson, 1985, p. 208). If redlining practices originated within the nation's financial institutions, the federal government sanctioned and reinforced such discriminatory practices at a critical time in the history of suburban development. The official stance of the federal government has changed in subsequent decades, but the patterns established by these policies have proven to be difficult to alter.

During the prosperous postwar years of the 1950s and 1960s urban redevelopment strategies were shaped by public-private partnerships. But the private partner dominated as the public sector's role consisted principally of "preparing the ground for capital." Spatially, the focus was on downtown and the suburbs. Socially, the dominant feature was the creation and reinforcement of racially discriminatory dual housing markets and homogeneous urban and suburban communities. These basic patterns have persisted in subsequent years when the national economy was not so favorable.

PARTNERSHIPS IN AN AGE OF DECLINE

The celebrated partnerships of the 1980s reflect an emerging effort to undermine the public sector, particularly the social safety net it has provided, and to reaffirm the "privileged position of business" (Lindblom, 1977) in the face of declining profitability brought on by globalization of the U.S. economy and its declining position in that changing marketplace. Government has a role, but again it is a subordinate one. As Bluestone and Harrison (1988) recently argued:

> Leaders may call these deals "public-private partnerships" and attempt to fold them under the ideological umbrella of laissez-faire. But they must be seen for what they really are: the re-allocation of public resources to fit a new agenda. That agenda is no longer redistribution, or even economic growth as conventionally defined. Rather, that agenda entails nothing less than the restructuring of the relations of production and the balance of power in the American

economy. In pursuit of these dubious goals, the public sector continues to play a crucial role. (pp. 107-108)

Global domination by the U.S. economy peaked roughly 25 years following the conclusion of World War II. After more than two decades of substantial economic growth subsequent to the war, international competition, particularly from Japan and West Germany but also from several Third World countries, began to challenge the U.S. position as productivity and profitability at home began to decline (Bluestone & Harrison, 1988; Bowles, Gordon, & Weisskopf, 1983; Reich, 1983). As both a cause and effect of the general decline beginning in the late 1960s and early 1970 the U.S. economy experienced significant shifts out of manufacturing and into service industries. Between 1970 and 1987 the U.S. economy lost 1.9 million manufacturing jobs and gained 13.9 million in the service sector (Mishel & Simon, 1988, p. 25). Perhaps even more important than the overall trajectory of decline has been the response to these developments on the part of corporate America and its partners in government. Such economic and political restructuring provided the context that has shaped the spatial development of cities and, in turn, the quality of life in urban America.

Between 1960 and 1980 the U.S. share of the world's economic output declined from 35% to 22% (Reich, 1987, p. 44). As profitability began to decline, U.S. corporations responded with an array of tactics aimed at generating short-term profits at the expense of long-term productivity (Hayes & Abernathy, 1980).

Rather than directing investment into manufacturing plants and equipment or research and development to improve the productivity of U.S. industry, corporate America pursued what Robert B. Reich labeled "paper entrepreneurialism" (Reich, 1983, pp. 140-172). That is, capital was expended on mergers and acquisitions, speculative real estate ventures, and other investments in which "some money will change hands, and no new wealth will be created" (Reich, 1983, p. 157). Rather than strategic planning for long-term productivity growth, the pursuit of short-term gain has been the objective.

Reducing labor costs has constituted a second component of an overall strategy aimed at short-term profitability. A number of tactics have been utilized to reduce the wage bill including decentralizing and globalizing production, expanding part-time work at the expense of full-time positions, contracting out work from union to non-union shops, aggressively fighting union organizing campaigns, implementing two-tiered wage scales, and outright demands for wage concessions. Rather than viewing human capital as a resource in which to invest to secure productivity in the long run, labor has increasingly been viewed as a cost of production to be minimized in the interests of short-term profitability (Bluestone & Harrison, 1988).

If production has been conceded by corporate America, control has not. Administration and a range of professional services have been consolidated and have grown considerably in recent years. If steel, automobile, and electronics production has shifted overseas, legal and accounting—along with other financial and related services—have expanded. Other service industries that have also grown include health care, state and local government, and personal services. Such developments lead some observers to dismiss the significance of a decline in manufacturing and celebrate the emergence of a postindustrial society (Becker, 1986; Bell, 1973). Yet at least half of those jobs in service industries are dependent on manufacturing production, though not necessarily production within the United States. Service and manufacturing are clearly linked; one cannot supplant the other. The health of both manufacturing and services depend on their mutual development. A service economy, without a manufacturing base to service, is proving to be a prescription for overall economic decline within those communities losing their industrial base (Cohen & Zysman, 1987).

True to the spirit of privatism, government has nurtured these developments through various forms of assistance to the private sector. Federal tax laws encourage investment in new facilities, particularly overseas, rather than reinvestment in older but still usable equipment, thus exacerbating the velocity of capital mobility (Bluestone & Harrison, 1982). State and local governments have offered their own inducements to encourage the pirating of employers in all industries ranging from heavy manufacturing to religious organizations (Eisinger, 1988; Goodman, 1979). Further inducements have been offered to the private sector through reductions in various regulatory functions of government. Civil rights, labor law, occupational health and safety rules, and environmental protection were enforced less aggressively in the 1980s than had been the case in the immediately preceding decades (Chambers, 1987; Taylor, 1989). If the expansion of such financial incentives and reductions in regulatory activity were initially justified in terms of the public benefits that would accrue from a revitalized private sector, in recent years unbridled competition and minimal government have become their own justification and not simply means to some other end (Bender, 1983; Connolly, 1983; Smith & Judd, 1984).

The impact of these structural developments is clearly visible on the spatial development of American cities. Accommodating these national and international trends, local partnerships have nurtured downtown development to service the growing service economy. If steel is no longer produced in Pittsburgh, the Golden Triangle has risen as the city's major employers now include financial, educational, and health care institutions (Sbragia, 1989). If auto workers have lost jobs by the thousands in Detroit, the Renaissance Center, a major medical center, and the Joe Louis Sports Arena have been built downtown (Darden, Hill, Thomas, & Thomas, 1987). Most major breweries have left Milwaukee, but the Grand Avenue Shopping Mall, several office buildings for legal, financial, and insurance companies, a new Performing Arts Center, and the Bradley Center housing the professional basketball Milwaukee Bucks are growing up in the central business district (Norman, 1989). With the U.S. economy deindustrializing and corporations consolidating administrative functions, downtown development to accommodate these changes is booming. These initiatives are more ambitious than urban renewal efforts that focused on rescuing downtown real estate, but many of the actors are the same and the fundamental relationships between the public and private entities prevail. In city after city such developments are initiated by the private side of local partnerships, usually with substantial public economic development assistance in the forms of UDAGs, IRBs, and other subsidies.

As cities increasingly become centers of administration, they experience an influx of relatively high-paid professional workers, the majority of whom are suburban residents (Levine, 1989, p. 26). Despite some pockets of gentrification, most of the increasing demand for housing for such workers has been in the suburbs. Retail and commercial businesses have expanded into the suburbs to service that growing population. To the extent that metropolitan areas have experienced an expansion of existing manufacturing facilities or have attracted new facilities, this growth has also disproportionately gone to the suburbs (Squires, Bennett, McCourt, & Nyden, 1987; White, Reynolds, McMahon, & Paetsch, 1989). Extending a trend that goes back before the war years, suburban communities have continued to grow.

The city of Chicago, often labeled the prototypical American industrial city, is also illustrative of the postindustrial trends. Between 1979 and 1987 downtown investment exceeded $6 billion as parking lots and skid row hotels have been replaced with office towers, up-scale restaurants and shops, and luxury housing (Schmidt, 1987). Yet overall during the postwar decades, manufacturing employment in the city has been cut in half while it tripled in the suburban ring. Total employment in the city of Chicago in the 1970s dropped by 14% while it increased by almost 45% in the suburbs (Squires et al., 1987). Continuation of downtown and suburban growth coupled with the decline of urban communities in between led *Chicago Tribune* columnist Clarence Page to describe his city as having a "dumbbell economy" (personal communication, March 19, 1987).

Throughout urban America, the rise of service industry jobs has fueled downtown and suburban development while the loss of manufacturing jobs has devastated blue-collar urban communities. Such uneven development is not simply the logical or natural outcome of impersonal market forces. The "supply-side" revolution at the federal level with the concomitant paper entrepreneurialism in private industry, the array of subsidies offered by state and local governments, and other forms of public intervention into the workings of the economy and the spatial development of cities, reveal the centrality of politics. As Mollenkopf concluded in reference to the postindustrial transformation of the largest central cities in the United States, "while its origins may be found in economic forces, federal urban development programs and the local progrowth coalitions which implemented them have magnified and channeled those economic forces" (Mollenkopf, 1983, p. 19). Uneven development therefore reflects conscious decisions made in both the public and private sectors in accordance with the logic of privatism, to further certain interests at the expense of others. Ideology has remained very much alive. Consequently, serious social costs have been paid.

Many of the social costs of both sudden economic decline and dramatic growth have been fully documented. As indicated above they include a range of economic and social strains for families, mental and physical health difficulties for current and former employees, fiscal crises for cities, and a range of environmental and community development problems. Among the more intangible yet clearly most consequential costs have been a reduction in the income of the average family and increasing inequality among wage earners and their families. Uneven economic and spatial development of cities has yielded unequal access to income and wealth for city residents.

For approximately 30 years after World War II family income increased and the degree of income inequality remained fairly constant. These trends turned around in the mid 1970s. Between 1977 and 1988 the vast majority of Americans experienced a decline in the buying power of their family incomes. Families in the lower 80% of the income distribution (four out of five families) were able to purchase fewer goods with their incomes by the end of the 1980s than they were able to do just 10 years earlier. The most severely affected were those in the bottom 10th who experienced a drop of 14.8%. Only those in the top 10th experienced a significant increase that, for them, was 16.5%. (Families in the 9th decile experienced a 1.0% increase). Most of this increase went to families in the upper 1% who enjoyed a gain of 49.8%. While GNP grew during these years and the purchasing power of the average family income increased by 2.2%, the top 20% received all of the net increase and more, reflecting the increasing inequality. Consequently, when adjusted for inflation, the lower 80% experienced a net decrease in the purchasing power of their family incomes (Gottdiener, 1990; Levy, 1987; Mishel & Simon, 1988, p. 6).

This growing inequality in the nation's income distribution reflects two basic trends. First is the shift from relatively high-paid manufacturing positions to lower paid service jobs. While service sector jobs include some highly paid professional positions, the vast majority of service jobs are low-paid, unskilled jobs. To illustrate, the Bureau of Labor Statistics projects an increase of approximately 250,000 computer systems analysts between 1986 and 2000 but more than 2.5 million jobs for waiters, waitresses, chambermaids and doormen, clerks, and custodians. There have also been income declines within industrial sectors reflecting the second trend, noted above, which is increasingly successful efforts by U.S. corporations to reduce the wage bill (Bluestone & Harrison, 1988).

Perhaps more problematic has been the growing racial gap. Racial disparities did decline in the first two decades following the war. Between 1947 and 1971 median black family income rose gradually from 51% to 61% of the white median. It fluctuated for a few years, reaching 61% again in 1975 but dropping consistently to 56% in 1987 (U.S. Bureau of the Census, 1976, 1980 [Table H7], 1989). For black men between the ages of 25 and 64 the gap improved between 1960 and 1980 from 49% to 64%, but dropped to 62% by 1987 (Jaynes & Williams, 1989, p. 28). Within cities, and particular big cities, the racial gaps have grown larger. Between 1968 and 1986 black median family income in metropolitan areas dropped from 63.7% to 57.4% of the white median. And in metropolitan areas with more than one million people, the ratio within the central city dropped from 69.7% to 59.0% (Squires, 1991).

Racial disparities in family wealth are even more dramatic. The median wealth of black households is 9% of the white household median. At each level of income and educational attainment, blacks control far fewer assets than do whites. Among those with monthly incomes below $900, black net worth is 1% that of whites with similar incomes. Education helps but does not close the gap. Among college educated householders 35 years of age or less with incomes of more than $48,000 annually, black net worth is 93% that of whites (Jaynes & Williams, 1989, pp. 276, 292).

Not only are blacks and whites separated economically, but racial segregation in the nation's housing markets persists. During the 1970s the degree of racial isolation in the nation's major cities remained virtually unchanged according to several statistical measures, leading two University of Chicago sociologists to identify 10 cities as "hypersegregated" (Massey & Denton, 1987, 1989). The degree of segregation differs little if at all by income for racial minorities. Among the consequences are unequal access to areas where jobs are being created and inequitable distribution of public services including education for minorities, and heightened racial tensions and conflicts for all city residents (Orfield, 1988).

Not surprisingly, it is predominantly black neighborhoods that have been most adversely effected by the uneven development of U.S. cities. For example, in Chicago between 1963 and 1977 the city experienced a 29% job loss but predominantly black communities lost 45% of all jobs. The increasing incidence of crime, drug abuse, teenage pregnancy, school dropout rates, and other indicators of so-called underclass behavior are clearly linked to the deindustrialization and disinvestment of city neighborhoods outside the central business district (Harris & Wilkins, 1988; Wilson, 1987).

Uneven structural and spatial development of cities adversely affects racial minorities. But racial inequalities in U.S. cities are not simply artifacts of those structural and spatial developments. Racism has its own dynamic. Blacks who have earned all the trappings of middle-class life in terms of a professional occupation, four-bedroom house, and designer cloths are still routinely subject to demeaning behavior that takes such forms as name calling on the streets by anonymous passers-by, discourteous service in restaurants and stores, and harassment on the part of police, all simply because of their race (Feagin, undated). Racially motivated violence in Bensenhurst and Howard Beach, Sambo parties and other racially derogatory behavior on several college campuses, and letter bombings of civil rights lawyers and judges confirm the continuity of vicious racism. The dynamics of class and race remain very difficult to disentangle, but the effects of both are all too real in urban America.

ALTERNATIVES TO THE PURSUIT OF THE PRIVATE CITY

Privatism and the policies that flow logically from that ideology have benefited those shaping redevelopment policy, including members of most public-private partnerships. But these policies have not stimulated redevelopment of cities generally. Structural, spatial, and social imbalances remain and are reinforced by the dynamics of privatism. To address the well-known social problems of urban America successfully, policies must be responsive to the structural and spatial forces impinging on cities. At least fragmented challenges to privatism have emerged in local redevelopment struggles in recent years. Alternative conceptions of development, the nature of city life, and human relations in general have been articulated and have had some impact on redevelopment efforts.

In several cities community groups have organized, and in some cases captured the mayor's office, in efforts to pursue more balanced redevelopment policies (Clavel, 1986). Explicitly viewing the city in terms of its use value rather than as a profit center for the local growth machine, initiatives have been launched to democratize the redevelopment process and to assure more equitable outcomes of redevelopment policy. Among the specific ingredients of this somewhat inchoate challenge to privatism are programs to retain and attract diverse industries including manufacturing, targeting of initiatives to those neighborhoods and population groups most in need, human capital development, and other public investments in the infrastructure of cities. A critical dimension of many of these programs is a conscious effort to bring neighborhood groups and residents, long victimized by uneven development, into the planning and implementation process as integral parts of urban partnerships.

When Harold Washington was elected mayor of Chicago in 1983, he launched a redevelopment plan that incorporated several of these components. The planning actually began during the campaign when people from various racial groups, economic classes, and geographic areas were brought together to identify goals and policies to achieve them under a Washington administration. Shortly after the election Washington released *Chicago Works Together: Chicago Development Plan 1984,* which reflected that involvement. Explicitly advocating a strategic approach to pursuing development with equity, the plan articulated five major goals: increased job opportunities for Chicagoans; balanced growth; neighborhood development via partnerships and coordinated investment; enhanced public participation in decision making; and pursuit of a regional, state, and national legislative agenda (*Chicago Works Together,* 1984, p. 1). As development initiatives proceeded under Washington, strategic plans were implemented that involved industrial and geographic sector-specific approaches to retain manufacturing and regenerate older neighborhoods, affirmative action plans to bring more minorities and women into city government as employees and as city contractors, provision of business incentives that were conditioned on locational choices and other public needs, and a planning process that involved community groups, public officials, and private industry.

Specific tactics included funding the Midwest Center for Labor Research to create an early warning system for the purposes of identifying potential plant closings and where feasible, interventions that would forestall the closing. Linked development programs were negotiated with specific developers to spread the benefits of downtown development. Planned manufacturing district legislation was enacted to control conversion of industrial zones to commercial and residential purposes, thus retaining some manufacturing jobs that would otherwise be lost. As the widely publicized "Council Wars" attested, Washington encountered strong resistance to many of his proposals (Bennett, 1988). The efforts of his administration demonstrated, however, that uneven urban development was not simply the outcome of natural or neutral market forces. Politics, including the decisions of public officials, mattered and those decisions under Washington were responsive to both public need and market signals (Giloth & Betancur, 1988; Mier, 1989; Mier, Moe, & Sherr, 1986).

In 1983 Boston also held a significant mayoral election. At the height of the Massachusetts miracle the city's economy was prospering and Raymond L. Flynn was elected with a mandate to "share the prosperity." Several policies have been implemented in order to do so.

Boston's strong real estate market in the early 1980s led to a shortage of low- and middle-income housing. Flynn played a central role in the implementation of a linkage program that took effect one month before he was elected. Under the linkage program a fee was levied on downtown development projects to assist construction of housing for the city's low- and middle-income residents. Shortly after taking office, the Flynn administration negotiated inclusionary zoning agreements with individual housing developers to provide below-market rate units in their housing developments or to pay an "in lieu of" fee into the linkage fund. To further alleviate the housing shortage, in 1983 the Boston Housing Partnership was formed to assist community development corporations in rehabilitating and managing housing units in their neighborhoods. The partnership's board includes executives from leading banks, utility companies, and insurance firms; city and state housing officials; and directors of local community development corporations.

Boston also established a residents job policy under which developers and employers are required to target city residents, minorities, and women for construction jobs and in the permanent jobs created by these developments. These commitments hold for publicly subsidized developments and, in an agreement reached by the mayor's office, the Greater Boston Real Estate

Board, the Buildings Trade Council, and leaders of the city's minority community, for private developments as well.

The Boston Compact represents another creative partnership in that city. Under this program the public schools agreed to make commitments to improve the schools' performance in return for the business community's agreement to give hiring preferences to their graduates. Schools have designed programs to encourage students to stay in school, develop their academic abilities, and learn job readiness skills. Several local employers, including members of the Vault, have agreed to provide jobs paying more than the minimum wage and financial assistance for college tuition to students who succeed in the public schools.

As in Chicago, the Flynn administration in Boston has consciously pursued balanced development and efforts to bring previously disenfranchised groups into the development process. The specific focus has been on housing and jobs, but the broader objective has been to share the benefits of development generally throughout the city (Dreier, 1989).

The Community Reinvestment Act (CRA) passed by Congress in 1977 has led to partnerships for urban reinvestment in cities across the nation. The CRA requires federally regulated banks and savings and loans to assess and be responsive to the credit needs of their service areas. Failure to do so can result in lenders being denied charters, new branches, or other corporate changes they intend to make. Neighborhood groups can challenge lenders' applications for such business operations with federal regulators, thus providing lenders with incentives to meet their CRA obligations (Potomac Institute, 1980). Subsequently, community groups and lenders have negotiated CRA agreements in more than 125 cities totaling approximately $6 billion in neighborhood reinvestment (Bradford, 1989). Examples include a $100-million loan pool created by 46 California banks to finance low-income housing, a $200-million commitment by Chase Manhatten Bank of New York for a community development fund, and a $245-million agreement negotiated by Chicago housing and neighborhood groups with several Chicago area lenders for various housing and business development projects (Guskind, 1989).

A unique lending partnership was created in Milwaukee in 1989. In response to the 1989 study finding Milwaukee to have the nation's highest racial disparity in mortgage loan rejection rates, the city's Democratic Mayor and Republican Governor created a committee to find ways to increase lending in the city's minority community. The Fair Lending Action Committee (FLAC) (1989) included lenders, lending regulators, real estate agents, community organizers, civil rights leaders, a city alderman, and others. An ambitious set of recommendations was unanimously agreed to in its report *Equal Access to Mortgage Lending: The Milwaukee Plan.* The key recommendation in the report was that area lenders would direct 13% of all residential, commercial real estate, and business loans to racial minorities by 1991. (After much debate the 13% figure was agreed to because that was the current minority representation in the population of the four-county Milwaukee metropolitan area.) Several low-interest loan programs were proposed to be financed and administered by lenders, city officials, and neighborhood groups. Fair housing training programs were recommended for all segments of the housing industry including lenders, real estate agents, insurers, and appraisers. The lending community was advised to provide $75,000 to support housing counseling centers that assist first-time home buyers. The city, county, and state were called upon to consider a linked deposit program to assure that public funds would go to those lenders responsive to the credit needs of the entire community. Specific recommendations were made to increase minority employment in the housing industry. And a permanent FLAC was called for to monitor progress in implementing the report's recommendations. The report concluded:

> There is a racial gap in mortgage lending in the Milwaukee metropolitan area. Implementation of these recommendations will be a major step in eliminating that gap. The Fair Lending Action Committee constitutes a partnership that is committed to the realization of fair lending and the availability of adequate mortgage loans and finance capital for all segments of the Milwaukee community. Building on the relationships that have been established among lenders, public officials, and community groups, neighborhood revitalization throughout the city and prosperity throughout the entire metropolitan area can and will be achieved. (Fair Lending Action Committee, 1989: 14)

At the press conference releasing the report, Governor Tommy G. Thompson "eloquently" stated, "Neither I nor the Mayor are the kind of guys who commission reports only to see them collect dust." The lenders indicated their own institutions and the local trade associations would support the report and implement its recommendations. Representatives of community groups, whose own reports have gathered much dust on bureaucrats' book cases, nodded approvingly in an expression of most cautious optimism.

Release of the report concluded what had been several months of contentious debates. The unanimous support for the report expressed by committee members at the conclusion did not negate the differences of opinion that prevailed or the fact that compromises were made in the interest of a show of unity. Yet the very existence of this wide-ranging report offers some additional hope for revitalization in Milwaukee. What remains to be seen is the extent of implementation.

These diverse initiatives are illustrative of experiments being launched in small towns and large cities in all regions of the United States. While they constitute an array of programs addressing a variety of problems, there are important underlying commonalities. They are responsive to the structural and spatial underpinnings of critical urban social problems. They are premised on a commitment to growth with equity; the notion that economic productivity and social justice can be mutually reinforcing. And the objective is to make cities more liveable, not just more profitable. A more progressive city is certainly not inevitable, but these efforts are vivid reminders that the major impediments have as much to do with politics as markets.

BEYOND LAISSEZ-FAIRE?

The trajectory of future redevelopment activity is blurred. The ideology of privatism is being challenged. Experiments with more progressive policies have occurred. But no linear path in the overall direction of public-private partnerships in particular or urban redevelopment in general has emerged. Harold Washington was soon followed by a Daley in Chicago. Boston's economy in the early 1990s does not look as promising as it did in the early 1980s and the demand for more incentives to the business community is getting louder in the wake of the Massachusetts miracle (personal communication, Peter Dreier of the Boston Redevelopment Authority, January 6, 1990). Milwaukee's mayor frequently expresses concern about the local business climate as civil rights groups challenge him to respond to the city's racial problems. Redevelopment remains a highly contentious political matter.

The grip of privatism has waned since the height of the Reagan years. HUD abuses, the savings and loan bailout, insider trading scandals, and other manifestations of the excesses of the pursuit of personal wealth serve as reminders of the importance of a public sector role beyond subsidization of private capital accumulation. Experiments in strategic planning to achieve balanced growth in Chicago, to share the prosperity in Boston, and to expand memberships in partnerships in Milwaukee and elsewhere demonstrate the capacity to conceive a different image of the city and the ability to implement programs in hopes of realizing that image. Yet as Warner concluded, "The quality which above all else characterizes our urban inheritance is privatism"

(1987, p. 202). For better or for worse, that remains the bedrock on which future plans will be built.

REFERENCES

Barnekov, T., Boyle, R., and Rich, D. (1989). *Privatism and urban policy in Britain and the United States.* New York: Oxford University Press.

Bennett, L. (1988). Harold Washington's Chicago: Placing a progressive city administration in context. *Social Policy, 19*(2), 22-28.

Becker, G. (1986, January 27). The prophets of doom have a dismal record. *Business Week,* p. 22.

Bell, D. (1960). *The end of ideology.* New York: Free Press.

Bell, D. (1973). *The coming of post-industrial society: A venture of social forecasting.* New York: Basic Books.

Bender, T. (1983). The end of the city? *democracy, 3* (Winter), 8-20.

Bluestone, B., & Harrison, B. (1982). *The deindustrialization of America: Plant closings, community abandonment, and the dismantling of basic industry.* New York: Basic Books.

Bluestone, B., & Harrison, B. (1988). *The great u-turn: Corporate restructuring and the polarizing of America.* New York: Basic Books.

Bowles, S., Gordon, D. M., & Weisskopf, T. E. (1983). *Beyond the waste land: A democratic alternative to economic decline.* Garden City, NY: Anchor Press/Doubleday.

Bradford, C. (1989). Reinvestment: The quiet revolution. *The Neighborhood Works, 12*(4), 1, 22-26.

Center for Community Change. (1989). *Bright promises, questionable results: An examination of how well three government subsidy programs created jobs.* Washington, DC: Center for Community Change.

Chambers, J. L. (1987). The law and black Americans: Retreat from civil rights. In J. Dewart (Ed.), *The state of black America 1987* (pp. 18-30). New York: National Urban League.

Chicago works together: Chicago development plan 1984. (1984). City of Chicago.

Clavel, P. (1986). *The progressive city: Planning and participation, 1969-1984.* New Brunswick, NJ: Rutgers University Press.

Cohen, S. S., & Zysman, J. (1987). *Manufacturing matters: The myth of the post-industrial economy.* New York: Basic Books.

Connolly, W. E. (1983). Progress, growth, and pessimism in America. *democracy, 3* (Fall): 22-31.

Darden, J., Hill, R. C., Thomas, J., & Thomas, R. (1987). *Detroit: Race and uneven development.* Philadelphia: Temple University Press.

Davis, P. (1986). *Public-private partnerships: Improving urban life.* New York: Academy of Political Science.

Dreier, P. (1989). Economic growth and economic justice in Boston: Populist housing and jobs policies. In G. D. Squires (Ed.), *Unequal partnerships: The political economy of urban redevelopment in postwar America* (pp. 35-58). New Brunswick, NJ: Rutgers University Press.

Dreier, P., Schwartz, D. C., & Greiner, A. (1988). What every business can do about housing. *Harvard Business Review, 66*(5), 52-61.

Eisinger, P. K. (1988). *The rise of the entrepreneurial state: State and local economic development policy in the United States.* Madison: University of Wisconsin Press.

Fair Lending Action Committee. (1989). *Equal access to mortgage lending: The Milwaukee plan.* Report to Mayor John Norquist and Governor Tommy G. Thompson (October).

Feagin, J. R. (1988). *Free Enterprise City: Houston in political and economic perspective.* New Brunswick, NJ: Rutgers University Press.

Feagin, J. R. (undated). The continuing significance of race: The black middle class in public places. Unpublished manuscript.

Friedland, R. (1983). *Power and crisis in the city: Corporations, unions and urban policy.* New York: Schocken.

Friedman, L. M. (1968). *Government and slum housing: A century of frustration.* Chicago: Rand McNally.

Giloth, R. & Betancur, J. (1988). Where downtown meets neighborhood: Industrial displacement in Chicago, 1978-1987. *Journal of the American Planning Association, 54*(3) 279-290.

Goodman, R. (1979). *The last entrepreneurs: America's regional wars for jobs and dollars.* Boston: South End Press.

Gottdiener, M. (1986). Retrospect and prospect in urban crisis theory. In M. Gottdiener (Ed.), *Cities in stress: A new look at the urban crisis* (pp. 277-291). Beverly Hills, CA: Sage.

Gottdiener, M. (1990). Crisis theory and state-financed capital: The new conjuncture in the USA. *International Journal of Urban and Regional Research, 14*(3), 383-404.

Greider, W. (1978). Detroit's streetwise mayor plays key role in city's turnaround. *Cleveland Plain Dealer* (July 3), cited in T. Swanstrom. (1985). *The crisis of growth politics: Cleveland, Kucinich, and the challenge of urban populism.* Philadelphia: Temple University Press.

Guskind, R. (1989). Thin red line. *National Journal, 21*(43), 2639-2643.

Harris, F., & Wilkins, R. W. (Eds.). (1988). *Quiet riots: Race and poverty in the United States.* New York: Pantheon.

Hartman, C. (1974). *Yerba Buena: Land grab and community resistance in San Francisco.* San Francisco: Glide.

Hayes, R. H., & Abernathy, W. J. (1980, July/August). Managing our way to economic decline. *Harvard Business Review, 58,* 67-77.

Hays, R. A. (1985). *The federal government & urban housing: Ideology and change in public policy.* Albany: SUNY Press.

Jackson, K. T. (1985). *Crabgrass frontier: The suburbanization of the United States.* New York: Oxford University Press.

Jaynes, G. D., & Williams, R. M. (Eds.). (1989). *A common destiny: Blacks and American society.* Washington, DC: National Academy Press.

Kasarda, J. (1988). Economic restructuring and America's urban dilemma. In M. Dogan & J. Kasarda (Eds.), *The metropolis era: Vol. 1. A world of giant cities* (pp. 56-84). Newbury Park: Sage.

Krumholz, N. (1984). Recovery of cities: An alternate view. In P. R. Porter & D. Sweet (Eds.), *Rebuilding America's cities: Roads to recovery* (pp. 173-192). New Brunswick, NJ: Center for Urban Policy Research.

Langton, S. (1983). Public-private partnerships: Hope or hoax? *National Civic Review, 72* (May), 256-261.

Levine, M. V. (1987). Downtown redevelopment as an urban growth strategy: A critical appraisal of the Baltimore Rennaissance. *Journal of Urban Affairs, 9*(2), 103-123.

Levine, M. V. (1989). The politics of partnership: Urban redevelopment since 1945. In G. D. Squires (Ed.), *Unequal partnerships: The political economy of urban redevelopment in postwar America* (pp. 12-34). New Brunswick, NJ: Rutgers University Press.

Levy, F. (1987). *Dollars and dreams: The changing American income distribution.* New York: Russell Sage.

Lindblom, C. E. (1977). *Politics and markets: The world's political-economic systems.* New York: Basic Books.

Logan, J. R., & Molotch, H. L. (1987). *Urban fortunes: The political economy of place.* Berkeley: University of California Press.

Massey, D. S., & Denton, N. A. (1987). Trends in the residential segregation of blacks, Hispanics, and Asians. *American Sociological Review, 52*(6), 802-825.

Massey, D. S., & Denton, N. A. (1989). Hypersegregation in U.S. metropolitan areas: Black and Hispanic segregation along five dimensions. *Demography, 26*(3), 373-391.

McKenzie, R. (1979). *Restrictions on business mobility: A study in political rhetoric and economic reality.* Washington, DC: American Enterprise Institute.

Mier, R. (1989). Neighborhood and region: An experiential basis for understanding. *Economic Development Quarterly, 3*(2), 169-174.

Mier, R., Moe, K. J., & Sherr, I. (1986). Strategic planning and the pursuit of reform, economic development, and equity. *Journal of the American Planning Association, 52*(3), 299-309.

Mishel, L., & Simon, J. (1988). *The state of working America.* Washington, DC: Economic Policy Institute.

Mollenkopf, J. H. (1983). *The contested city.* Princeton, NJ: Princeton University Press.

Norman, J. (1989). Congenial Milwaukee: A segregated city. In G. D. Squires (Ed.), *Unequal partnerships: The political economy of urban redevelopment in postwar America* (pp. 178-201). New Brunswick, NJ: Rutgers University Press.

Orfield, G. (1988). Separate societies: Have the Kerner warnings come true? In F. R. Harris & W. Wilkins (Eds.), *Quiet riots: Race and poverty in the United States* (pp. 100-122). New York: Pantheon.

Peterson, G. & Lewis, C. (Ed.). (1986). *Reagan and the cities.* Washington, DC: The Urban Institute.

Peterson, P. E. (1981). *City limits.* Chicago: University of Chicago Press.

Porter, D. R. (1989). Balancing the interests in public/private partnerships. *Urban Land, 48*(5), 36-37.

Potomac Institute. (1980). *Lender's guide to fair mortgage policies.* Washington, DC: Author.

President's Commission for a National Agenda for the Eighties. (1980). *A national agenda for the eighties.* Washington, DC: Government Printing Office.

Reich, R. B. (1983). *The next American frontier.* New York: Times Books.

Reich, R. B. (1987). *Tales of a new America.* New York: Times Books.

Sbragia, A. (1989). The Pittsburgh model of economic development: Partnership, responsiveness, and indifference. In G. D. Squires (Ed.), *Unequal partnerships: The political economy of urban redevelopment in postwar America* (pp. 103-120). New Brunswick, NJ: Rutgers University Press.

Schmidt, W. (1987, October 11). U.S. downtowns: No longer downtrodden. *The New York Times.*

Schumpeter, J. (1942). *Capitalism, socialism, and democracy.* New York: Harper & Row.

Smith, M. P., & Judd, D. R. (1984). American cities: The production of ideology. In M. P. Smith & D. R. Judd (Eds.), *Cities in transformation: Class, capital, and the state* (pp. 177-196). Beverly Hills, CA: Sage.

Squires, G. D. (1991) Deindustrialization, economic democracy, and equal opportunity: The changing context of race relations in urban America. *Comparative Urban and Community Research, 3,* 188-215.

Squires, G. D., Bennett, L., McCourt, K., & Nyden, P. (1987). *Chicago: Race, class, and the response to urban decline.* Philadelphia: Temple University Press.

Stone, C. N. (1987). The study of the politics of urban development. In C. N. Stone & H. T. Sanders (Eds.), *The politics of urban development* (pp. 3-22). Lawrence: University of Kansas Press.

Taylor, W. L. (1989). Special report: Supreme Court decisions do grave damage to equal employment opportunity law. *Civil Rights Monitor,4*(2), 1-28.

Thomas, J. M. (1989). Detroit: The centrifugal city. In G. D. Squires (Ed.), *Unequal partnerships: The political economy of urban redevelopment in postwar America* (pp. 142-160). New Brunswick, NJ: Rutgers University Press.

U.S. Bureau of the Census. (1976). *The statistical history of the United States: From colonial times to the present.* New York: Basic Books.

U.S. Bureau of the Census. (1980). *Structural equipment and household characteristics of housing units.* Washington, DC: Government Printing Office.

U.S. Bureau of the Census. (1989). *Statistical abstract of the United States: 1989.* Washington, DC: Government Printing Office.

Warner, S. B., Jr. (1987). *The private city: Philadelphia in three periods of its growth.* Philadelphia: University of Pennsylvania Press.

White, S. B., Reynolds, P. D., McMahon, W., & Paetsch, J. (1989). *City and suburban impacts of industrial change in Milwaukee, 1978-87.* Milwaukee: University of Wisconsin-Milwaukee, The Urban Research Center.

Wilson, W. J. (1987). *The truly disadvantaged: The inner city, the underclass, and public policy.* Chicago: University of Chicago Press.

10

The Reshaping
of Urban Leadership in U.S. Cities:
A Regime Analysis

CLARENCE N. STONE
MARION E. ORR
DAVID IMBROSCIO

FOR MORE THAN 40 YEARS the top priority on the urban agenda in America has been the physical redevelopment of the city. Since World War II, urban leaders and national policy makers have adjusted to the changing social economy of the city by modifying land use in an effort to spur investment within downtown areas. The building of office towers, hotels, convention centers, exhibition halls, festival marketplaces, sports facilities, and transportation networks has been the predominant concern of city leaders. Although some of the entrepreneurs of redevelopment were appointed executives, many were elected. New Haven's Mayor Richard Lee became a celebrated figure in political science and journalism by demonstrating how a strong and durable coalition could be built around city rejuvenation, featuring a revitalized central business district.

Yet the kind of city renewal typified by New Haven has proved to be no cure-all. In a variety of cities, physical restructuring has done little to improve the lives of many urban dwellers. Nationwide, one in five children is born into poverty. In many large cities, the proportion is twice that figure.

In his survey of the postwar urban experience, Marc Levine (1989) maintains that "in city after city, redevelopment has been associated with a

AUTHORS' NOTE: Clarence N. Stone wishes to express appreciation to the College of Urban, Labor, and Metropolitan Affairs of Wayne State University for a research appointment that made the groundwork for this chapter possible.

tale of two cities:' pockets of revitalization surrounded by areas that experience growing hardship" (p. 25). Even efforts to preserve manufacturing jobs have had limited effect (Jones & Bachelor, 1986). And the more typical effort produces corporate-centered, service-oriented employment that either benefits surrounding suburban residents or offers city residents low-paying dead-end jobs (Fainstein & Fainstein, 1989). As a consequence, many inner-city neighborhoods have deteriorated and poverty proves stubbornly persistent (Kasarda, 1985; Wilson, 1987).

This pattern of emphasizing redevelopment and its having a disappointing impact on urban life has been known for a long time. So why have alternative strategies not been tried? If physical restructuring has not prevented social decay, why not pursue a "people approach"—why not pursue a policy of human-capital enhancement? For a time in the 1960s under the aegis of the Great Society, a few cities like Detroit under Mayor Cavanagh and New York under Mayor Lindsay made some moves in that direction, but diminishing federal funds quickly dissipated city hall support for priority attention to human-resource problems. And especially after New York City's fiscal crisis in the 1970s, the rush was on again to pursue physical restructuring as a way of protecting the tax base and avoiding deep service cutbacks. Still, we should not overlook the fact that human-capital issues such as educational reform enjoy recurring attention in campaign rhetoric, even though they seem never to move very high on the action agenda of city halls.

URBAN REGIMES

To understand why redevelopment is a durable priority while attention to human capital is so limited, we need to appreciate that policy making is not simply the product of deliberation and enactment. It is a matter that involves coalition building and assembling resources (Mollenkopf, 1983). For that reason, governing coalitions may differ substantially from winning electoral alliances (Ferman, 1985). Despite much talk about state autonomy (national and local), the authority of government has a weak writ, and that is especially the case for American cities. Urban governance necessitates strengthening what is formally and officially a weak capacity to make and carry out policies. Inevitably city officials seek alliances that enhance their ability to achieve visible policy results (Stone, 1980). Thus to understand policy making we need to consider how the limited resources commanded by public officials are *melded together* with those of private actors to produce a capacity to govern. The arrangements by which such governing coalitions are created can be called *regimes*—in the case of localities, urban regimes. Governance rests less on formal authority than on arrangements through which public officials and private interests create a complex system of cooperation.

Policy making is thus not simply a matter of choosing a reasonable course of action; it is shaped by the composition of the governing coalition, the terms that underlie the cooperation of coalition members with one another, and the resources they are capable of assembling. This means that regimes are unable to pursue any and all policy priorities equally well. Pursuing a policy priority such as human-capital development requires a regime whose members are not only supportive of that aim but also up to the task of bringing together the necessary resources.

Now we can suggest a line of explanation for the prevalence of physical restructuring as a policy priority over the pursuit of human-capital development. Possibly urban regimes devoted to and capable of pursuing redevelopment are easier to build than regimes devoted to and capable of pursuing human-capital development. It may well be that, other things being equal, what is easier to assemble is able to crowd out what is harder to construct, especially given the short-time frame that elected officials operate within. Differences in ease of regime organization may explain why, in urban politics, electoral alliances and campaign rhetoric are such poor predictors of subsequent policy activity.

Priorities, we are suggesting, grow out of the regime arrangements that are readily put together. Like an organization, a regime seems to discover "preferences through action more than it acts on the basis of preferences" (Kingdon, 1984, p. 89). What can be done easily thus holds an advantage over what can be done only with difficulty.

Urban research needs to explore the conditions underlying varying policy capacities and what their consequences are. For this inquiry, we offer the following as a guiding proposition: In order for a governing coalition to be viable, it must have a capacity to mobilize resources commensurate with the requirements of its main policy agenda. If a coalition cannot deliver on the agenda that brought it together, then the members will disengage, leaving the coalition open to reconstitution. On the other hand, doable actions help secure commitments and perhaps attract others with similar or consistent aims.

THE TASK OF REGIME BUILDING

We shall now turn to an examination of the task of regime building and what is involved. In this way we can see how regime capacity shapes policy activity. First, however, let us consider some alternative explanations about why redevelopment has enjoyed higher priority on the urban agenda than has human-capital enhancement. It is not the case that human-resource issues have been conflict laden while physical restructuring has enjoyed a largely controversy-free existence. Redevelopment has been accompanied by an enormous amount of community conflict, conflict in more places and extending over more years than what one would find in the human-capital arena. Furthermore, it is *not* the case that business favors only physical redevelopment and opposes human-capital efforts. Many business leaders have supported measures intended to promote an educated and skilled work force. And such a work force is almost certainly a contributor to a favorable climate for business investment and therefore a protector of the tax base as well. So the merits of the two sets of policies and sentiment about them offer no plausible answer to the differing fates of physical-redevelopment versus human-capital regimes.

Instead, let us turn to the nature of the regime itself. In international affairs, a regime is an informal system of cooperation based on the enforcement capacity of a hegemonic power or perhaps on norms adhered to voluntarily. In either case, cooperation rather than shortsighted pursuit of national interest is the rule (Krasner, 1983).

As it operates in the urban community, the regime can again be seen as a system of cooperation. What, then, facilitates cooperation around physical-redevelopment programs that is not present to the same degree in human-resources programs? There seems to be no hegemonic power, but one possible explanation is a privatist ideology (see M.V. Levine, 1989; Smith, 1988). Conceivably a shared belief in economic growth facilitates cooperation, but that line of argument does not account for the extensive conflict that has surrounded the physical restructuring of the city. Furthermore, it is a line of argument that assumes that general beliefs guide actions. Students of policy implementation long ago called that assumption into question. Pressman and Wildavsky (1984) found that subscription to a general view is a poor guide to specific behavior. Moreover, some students of cooperative behavior argue, as we have seen, that action shapes preferences more than preference shapes action, and especially if the preference is vague and general (Cohen & March, 1986; see also Perrow, 1986). Besides, if belief is the key factor, why would general sentiment in favor of human-capital development not lead to an ideology of opportunity-enhancement, that would then pave the way for cooperation around human-capital policies? The argument from belief is not persuasive. We must conclude that, standing by itself, a privatist ideology does not offer a very satisfactory answer to the question about the prevalence of redevelopment regimes.

Let us turn to another possibility. Physical restructuring, particularly when assisted by substantial federal and state grants, is characterized by an abundance of selective material incentives (see, for example, Caro, 1974). As a collective-action problem, redevelopment would be aided immensely by such incentives. And these incentives are useful not only for enlisting support, but also for disbanding or at least fragmenting opposition (see Jones & Bachelor, 1986). The availability of selective material incentives is, then, a plausible explanation for the viability of physical-redevelopment regimes, even in the face of significant resistance.

There is, however, another feature of such regimes we should give attention to. Unlike intensified development of human capital, physical restructuring requires no mass base of participation. It can be done by coordination at the top, coordination among peak organizations representing a few key sectors of the community (see especially Jones & Bachelor, 1986). By contrast, human-resource efforts require mobilization of a mass population *in addition to* peak-level coordination among sectors of the community. In short, physical restructuring constitutes an easier task of cooperation than does the development of human capital, and there are resources in the form of selective material incentives that are readily available.

For contrast, let us turn to what it would take to make a regime oriented toward human-resource development viable. The short answer is "a great deal." In particular the challenge of mass mobilization has to be faced, and such mobilization is both difficult to achieve and if successful potentially threatening to established power centers. For that reason, it is not a task that is intrinsically appealing to those who are officeholders.

Consider the matter of difficulty first. The mass mobilization needed is not a one-time show of support. Instead, it is a sustained involvement in seeing that human-capital opportunities are not only present on a wide scale, but also actively used. This means a regime operating at two levels: (1) a peak level where resources for human-capital programs are gathered, and (2) a mass level where people need to be motivated to enroll in the appropriate programs, make full use of the opportunities available, and provide active support for their continuation.

The two levels are integrally related. Without a substantial body of opportunities, motivation on a mass basis is not achievable. Without mass mobilization to keep the pressure on, peak-level efforts to provide resources and opportunities are not likely to be durable. But pressure is never welcomed. And those who are pressured see the potential for, but wish to avoid, destabilization in mass mobilization. What they want are supportive relationships that can be counted on, not arrangements in which support is always in question. That is one reason why mass mobilization, as something more unpredictable, is much harder to achieve than peak-level coordination. Mass participation in human-capital programs could presumably be stabilized over time and reciprocal understandings worked out, but it cannot be done by peak-level bargaining. If the mass is to be motivated, then it needs to be directly involved in the development of the arrangements. The organizational problem is therefore formidable.

The contrast between redevelopment regimes and human-capital regimes suggests that easier regime building tasks will prevail over more difficult ones, other things being equal. But, of course, other things are rarely equal. Thus it is particularly useful to consider the availability of resources to those who would build, lead, and maintain urban regimes. Let us return to the proposition offered earlier: *In order for a governing coalition to be viable, it must have a capacity to mobilize resources commensurate with the requirements of its main policy agenda.* What this proposition incorporates is an acknowledgment that variations in capacity to assemble and use resources are important. A more difficult organizational task can potentially crowd out an easier organizational task if proper and sufficient resources are available. Hence a caretaker regime is easier to organize than one devoted to restructuring urban land use. But in many cities redevelopment regimes replaced caretaker regimes, though not in every case (Sanders, 1987). And in a few largely middle-class communities progressive regimes replaced redevelopment or caretaker regimes (Clavel, 1985; DeLeon, 1990; Rosdil, 1989).

Regime-building, then, is not an autonomous process whereby in market-like fashion numerous shortsighted transactions produce a set of governing arrangements. Instead of such an invisible hand, we have a political process in which leadership may be exercised and purposive choices made regarding how to go about community governance. To be sure, no one's master plan is enacted; even purposive choices involve a limited understanding of what is possible and what the consequences of particular decisions are. But whether we are looking at the Daley machine in Chicago, the formation of a biracial alliance in Atlanta, Richard Lee's executive-centered coalition in New Haven, or other instances of regime building, some body of actors is engaged in a conscious and intentional process of shaping arrangements for the governance of the city.

Leadership enters the picture as a force that can broaden the vision of what is possible and increase understanding about consequences of decisions. But leadership does not operate in a completely open-ended context; what can be done is partly determined by what is already in place. Here we are talking about more than the particular features of a previous regime. We are directing attention to the broad structural features of the American political economy: (1) a large and varied private sector that not only controls most investment activity but also houses a lively array of associational life, and (2) governmental authority that relies more on inducing actions than it does on simply issuing commands—a feature that is even more pronounced at the local level than at higher levels of government in the United States.

What, then, are the ingredients for a process of regime formation in which there is electoral control of a weak system of governmental authority and private control of important material and organizational resources? Two considerations are especially important: (1) private resources are unevenly distributed; and (2) the motivation to participate and use those resources is also unevenly distributed. These two considerations can serve as elements in an explanation of regime formation. Our beginning point is one made earlier, namely that the task of governance is much more demanding than simply assuming control of public office. Regime formation therefore cannot be a mere matter of aggregating electoral support around a policy program. Instead, it is a matter of bringing private and public resources to bear in the production of a coherent policy effort; and, as indicated above, *the mobilization of resources must meet the demands of the policy agenda.* This stipulation underscores the importance of the distribution of resources and of the motivation to participate. To govern, a coalition must have at its disposal resources that are commensurate with the policy-making effort it is based on.

With these considerations in mind, we turn now to a quick overview of the experience in American cities. In doing so, we want to highlight the special difficulties of promoting human-capital development. Then we can consider more concretely what challenge urban leadership must meet if it is to succeed in altering the policy agenda so as to address more effectively the human-resource problems of the nation's cities.

URBAN REGIMES AND
THEIR POLICY AGENDAS

In the discussion that follows, we move from the types of regimes that involve the simplest organizational problems to those that involve the most complicated, and therefore from those that involve the least resource requirement to the greatest.

I. TYPE—CARETAKER AND EXCLUSIONARY

Main Task—Maintenance

Explanation. This type involves mainly the provision of routine services and beyond that needs only periodic popular approval at the ballot box. Exclusionary policies can entail a degree of regulation beyond the routine, but, like the simple caretaker orientation, call for no intensive motivation of a large number of actors and can therefore be satisfied with a relatively small resource base.

Discussion. Since a caretaker regime requires little to sustain itself, it is vulnerable only if a challenge coalition can create the resource base it needs. Should the caretaker stance be challenged, its protection needs only the mobilization of a majority vote—it is not dependent on an extensive and ongoing commitment of resources by participants. The motivational requirements are slight and can be met with relative ease on a mass basis by small property holders concerned over tax levels (see the discussion of Kalamazoo by Sanders, 1987). At the same time, caretaker government offers little advancement opportunity for ambitious elected or appointed officials. Adherence to the caretaker stance comes easiest to "provincial" officeholders, who are content with lives embedded in particular local communities.

Suburbs practicing economic and racial exclusion are quite similar in motivational demands. The task usually involves no great challenge and little policy innovation. Regulation of land use generally suffices, although the extent of governmental effort demanded varies somewhat with the location and the external constraints on the jurisdiction. Dearborn, Michigan, for example, has pursued a complex strategy of exclusion—one that entailed more than a minimal policy effort of large-lot zoning (see Darden, Hill, Thomas, & Thomas, 1987; Good, 1989). A once compliant Ford Motor Company provided a tax windfall that enabled the Orville Hubbard machine to become entrenched and practice its multifaceted strategy of exclusion. Like caretaker officials, Hubbard was very much a "provincial."

II. TYPE—BUSINESS-CENTERED ACTIVIST (REDEVELOPMENT)

Main Task—Coordination

Explanation. This type involves the *active* provision of subsidies and inducements to businesses to promote investment in development and redevelopment. Development is not a matter of isolated decisions about investment; rather, it usually involves a shared view that an area is suitable for risks and that a number of actors are willing to move in concert to upgrade, conserve, or transform it. Transportation and other forms of public infrastructure, including convention centers, aquariums, and a variety of support facilities, may be part of a planned inducement of private investment. Such a development policy typically calls for coordination among various institutional elites, and the policy needs to result in the generation of benefits that can be provided to well-placed subelites who can assist in conflict management.

Discussion.   Throughout the second half of the 20th century most central cities have been characterized by activist regimes, dedicated to a business-centered agenda of downtown revitalization. These are programs that are often controversial in their particulars, and on occasion result in electoral defeats. Although the policy task can be executed only with substantial private resources, it is one that generates an ample supply of selective material incentives. And these incentives are enormously helpful in coping with the complexity of joint action so central to the process of policy implementation (Pressman & Wildavsky, 1984).

With federal urban redevelopment policy from the beginning tilted toward reliance on private investment, business-centered governing coalitions have received strong external encouragement. However, business-centered coalitions heavily concerned with development are by no means dependent on external stimulation. These coalitions—termed aptly by Logan and Molotch (1987) as "growth machines"—epitomize the convergence of resources and motivation. Real estate and developer interests, financial institutions, newspapers, utilities, and other large downtown property holders have a major stake in the enhancement of land values in the central city; and they have the resources to devote to the cultivation of a favorable political climate as well as the technical expertise and financial capacity to promote development. Even so, their mutual collaboration depends on active encouragement. In many cities, downtown landholding interests have created organizations through which plans and actions can be coordinated (one of the earliest was Pittsburgh, see Lubove, 1969). Sometimes entrepreneurial public administrators, like Robert Moses or Edward Logue, have served as the principal coordinating force (see, for example, Caro, 1974). In other cases, it has been particular business figures, like David Rockefeller (Fainstein & Fainstein, 1989). And in a few instances it has been local foundations that have played the key coordinating role (Jones & Bachelor, 1986). In other instances, coordination is weak (see, for example, Lupsha, 1987).

Given public officials who want a record of visible accomplishment, the attractions of a business alliance are readily explained. Even when African-

American mayors come into office with platforms promising a program of expanded opportunities for lower class constituents, these mayors are often diverted into more narrow, business-supported policy efforts—these are the policy efforts that can attract the necessary private resources, and the business-centered coalition is the one that can be created quickly (Nelson, 1987; Preston, 1987; Reed, 1988; Rich, 1989). That these policy efforts generate substantial particular benefits means that ideological accord is not a requirement, and racial division can be bypassed. Thus even though the transaction costs for redevelopment policies in racially divided communities may be relatively high (C. H. Levine, 1974), a business-oriented regime's capacity to meet those cost is also substantial (Stone, 1989).

III. TYPE—MIDDLE-CLASS PROGRESSIVE

Main Task—Complex Regulation

Explanation. Middle-class progressives often favor such measures as historic preservation, environmental protection, affordable housing to preserve and promote population diversity, affirmative action in employment and business contracts, and linkage funds for various social purposes. These measures entail imposing restrictions on and exactions from development and other business interests. Exactions, especially, require that the source activities be encouraged or at least not discouraged. Hence progressive mandates involve monitoring the actions of institutional elites and calibrating inducements and sanctions to gain a suitable mix of activity and restriction.

Discussion. The few American communities that have relatively stable regimes organized around a progressive agenda are places like Burlington, Vermont, with a largely middle-class population (Clavel, 1986; Rosdil, 1989). Moreover, these are communities in which the middle class is organized independent of the sponsorship of corporate business. In short, there is a nonbusiness resource base that can be tapped to support the difficult task of putting into effect an array of progressive mandates. Some cities, such as Boston, Seattle, and San Francisco, have a mix of policy orientations that includes significant progressive measures. These, too, are cities with middle-class activists unattached to corporate business, and they are also cities with substantial research and intellectual capacity not tied to corporate business. Other resources also enter the picture. New York City's long-running progressive housing policy rested partly on the availability of financial resources from labor unions (Fainstein & Fainstein, 1989).

Although regulation does not require extensive participation by masses of people, progressive mandates often, in fact, rest on a base of active popular support. Because progressive mandates may involve significant tradeoffs, citizen participation is useful in informing citizens about the complexities of policy while keeping them committed to progressive goals. And if the referendum process is a keystone in regulation, as in San Francisco (DeLeon, 1990), then dependence on mass support is direct and central.

IV. TYPE—LOWER CLASS OPPORTUNITY EXPANSION

Main Task—Mobilization

Explanation. This is the area of urban policy that may well be least understood. We prefer to avoid the term *redistribution* because that term treats the lower class as simply claimants for greater service. The real point is to *expand* opportunities for the lower class through the enhancement of human capital and widened access to employment and ownership. Contrary to what the word *redistribution* suggests, the process need not be zero-sum; it can provide community-wide benefits. Certainly it is the case that business executives frequently endorse the idea of educational reform and enhanced human capital. Thus such specific activities as enriched education and job training, improved transportation access, and enlarged opportunities for employment and for business and home ownership often draw wide expressions of support.

While we should not ignore the considerable potential for social conflict in the particulars of how to achieve these goals, broad-scale community conflict is not the only obstacle. Motivation is a large consideration. Machiavelli captured the matter when he observed:

> nothing is more difficult than to handle, more doubtful of success, nor more dangerous to manage, than to put oneself at the head of introducing new orders. For the introducer has all those who benefit from the old orders as enemies, and he has lukewarm defenders in all those who might benefit from the new orders. This lukewarmness arises partly from fear of adversaries who have the laws on their side and partly from the incredulity of men, who do not believe in new things unless they come to have a firm experience of them. (1985, pp. 23-24)

The policy challenge is how to generate and maintain active support for "new things" before the masses have had an opportunity "to have a firm experience of them." If Machiavelli is right, the motivation to support comes after the fact, but political reality calls for active support as a precondition of basic policy change.

What might be entailed in a policy strategy directed at expanding lower class opportunities through the enhancement of human capital? A contrast with routine services might be illuminating. Schooling as a routine service calls for no early childhood education, no special efforts at increasing the school "readiness" of children, no extraordinary measures to increase motivation for those who have been socialized into an environment of restricted opportunity and low expectations. Education as a routine service simply accepts school performance as largely a product of social background; it entails no special efforts to enrich opportunities and raise expectations. The same could be said for job training and job placement, for transportation, and for home and business ownership—some set of opportunities are there for those who have the requisite credentials, connections, and personal aspirations.

Altering opportunities on a class basis calls for more than pep talks to individuals about working hard to get ahead. In the first place, the process requires that opportunities be real—that those who meet education or training requirements be offered decent jobs, not dead-end jobs with no future (Bernick, 1987). School compacts that guarantee jobs to high school graduates or that assure financial support for a college education are the kinds of practices that make opportunities real. A few individual opportunities or scattered chances to compete for a restricted set of positions are not enough.

The availability of the opportunities is only the first step. Lower class children also need to believe that the opportunities are real and that they are actually attainable. Given a background that encourages low expectations and cynicism about life chances, members of the lower class are likely to pursue opportunities only if they are encouraged and supported not only individually but also through their families and their peers. Put another way, changing conditions on paper is not enough. Previously conditioned expectations have to be altered. To do that, it is necessary to create a complex set of incentives that are extensive enough to affect class-wide views and that are intensive enough to sustain ongoing personal commitments to make use of expanded opportunities.

Discussion. Executing a program through which opportunities for the lower class would be expanded is clearly the most difficult policy effort that a coalition could undertake. The principal difficulty lies not in an absence of program ideas, but in providing continuing support to put these ideas into sustained practice and refine them as experience indicates. There is no readily available resource base for this effort. The urban lower class faces a paradox—it lacks the organization and other resources to make an opportunity-expanding regime feasible but without such a regime it faces continuing neglect in the policy choices that are made. The experience of Latino and African-American mayors in various cities is that electoral success is not enough to provide a foundation for class-inclusionary policy regimes. Hence, once in office, mayors tend to embrace what is most readily available—programs that provide expanded opportunities for the minority middle class while offering to the masses mainly the symbolic appeal of communal-group solidarity. The kind of large-scale campaign needed to expand opportunities on a class basis would be long and hard. As predicted in our guiding proposition, class-inclusionary regimes fail to move from rhetoric to reality because elected officials cannot mobilize resources commensurate with the requirements of such a policy agenda.

Does that mean that class-inclusionary regimes are completely unfeasible? Not necessarily. But it does suggest that the mobilization of the lower class in ways that are policy relevant—given the scope of the collective-action problem involved and the scarcity of private resources they command—seems most likely when external resources are available. Even with external resources, opportunity expansion is likely to be a long-range campaign, taking place in stages.

The motivational problem allows no easy solution. Sustained mass mobilization requires an orchestration of efforts by established institutions to see that the opportunities are there, and efforts within lower class circles to raise and act on expectations (cf. Hochschild, 1989; on the difficulties of mobilizing the lower class, see Henig, 1982). Elite-level coordination and the monitoring of selected community activities are thus not enough. The task is enormous, and its enormity may discourage public officials and others looking for quick and visible results from taking it on. Nevertheless, it is in order to be realistic about what the expansion of lower class opportunities calls for.

Whether there are more than fragmentary examples of opportunity-expanding arrangements in the United States is thus doubtful, but inquiry should be a research priority. The 1960s saw a few significant initiatives in this direction. Under Mayor Cavanagh, Detroit launched a broad effort in partnership with the federal government. Yet, in retrospect, the effort looks thin. Certainly it was weakened by the spread of federal money to a large number of other communities, and the diversion of federal resources and energy into the Vietnam war. Moreover, Detroit was hit by a devastating riot, and Cavanagh himself failed to expand or even sustain his own "model city" effort.

Under John Lindsay, New York City also made some preliminary efforts to mobilize the lower class and expand the opportunities available to them.

Although some organizational fruits of the earlier effort still remain, community conflicts left the Lindsay administration weakened in its second and final term (Fainstein & Fainstein, 1989). As in Detroit, we look back and see a halting and limited effort. In both instances there may, however, be hints of what is possible on a more lasting and extended basis.

Among various areas of policy, education is an arena in which there have been small-scale successes. But the leverage to expand these successes is often missing—especially when, as has apparently been the case in some large cities, no member of the Board of Education has children in the public schools. As Machiavelli suggests, reform requires not only that powerful opposition be overcome, but also that the backers of reform be motivated to press their case. When the most direct beneficiaries of reform are not even part of the governing coalition, it is hard to see how a cycle of cynicism and disappointment can be overcome.

Perhaps more substantive policy efforts have been made in some European cities. If so, it seems likely that resources from central governments were important ingredients. Certainly it is the case that in cities such as Sheffield, England—where there once was a substantial commitment to working-class mobilization—efforts have been undermined by changes in national policy (Seyd, 1990).

Still, even without substantial federal assistance—as seems likely to be the continuing case in the United States—there are a range of policy initiatives worth looking at: preschool and other educational programs that work, school compacts that guarantee employment and college opportunities, training and placement that lead to jobs that are not dead-ended, home- and business-ownership opportunities broader than those that are market-supplied, the creation and support of nonprofit and community-based organizations, and the reshaping of investment practices through the invocation of such laws as the Community Reinvestment Act. The base of support for these initiatives and the obstacles that surround their use are significant matters.

While regimes devoted to opportunity expansion for the lower class are more a logical extension of practice than a full reality, we should not ignore the possibilities suggested in concrete practice. Because actual regimes may be mixtures of the types described here, cities predominantly of one kind may display tendencies toward another type. Hence, small elements of an opportunity-expansion regime may be found in various cities. San Francisco's progressivism, for example, is not exclusively middle class (DeLeon, 1990). Chicago after Harold Washington's election provided wider opportunities than Chicago before Washington (Bennett, 1989). Equity or advocacy planning, as developed in Cleveland, provided an alternative set of priorities

(Swanstrom, 1985, pp. 114-116). Boston's Housing Partnership and the Boston Compact represent significant moves to open benefits to the lower class (Dreier, 1989). Although movement may have been small, each of these cities has edged toward a more class-inclusionary regime.

LEADERSHIP

Leadership is not a process of presiding over what is inevitable. It is about making something happen that, given the ordinary course of events, would not occur. For urban leaders to bring about the extraordinary result of a regime devoted to opportunity expansion for the lower class would entail overcoming the obstacles of an unequal distribution of resources and motivations. These obstacles manifest themselves in the form of (1) business control of capital wealth and exceptional influence over the tax base, (2) cities with weak authority and limited jurisdictions in ever-expanding metropolitan areas, and (3) social isolation and a scarcity of neighborhood-level organizations and institutions in lower class sections of the city.

A test of urban leadership is whether or not it can modify the impact of unequal resources and motivation by altering these three obstacles. Consider them in succession.

First, will business actively support only physical restructuring and divorce itself from responsibility for the social and human aspects of urban life? That is perhaps the path of least resistance. Leadership could, however, contribute to an altered impact by persuading corporate executives that the long-term interest of business as well as that of society more generally depends on human-capital development. Persuasion, in turn, could be aided by national legislation tying corporate privileges to the assumption of social responsibilities. The Community Reinvestment Act represents one step in that direction, but other measures could also be enacted, or, in lieu of enactments, business executives could sign compacts acknowledging their obligations to contribute to society as the source from which they derive their economic opportunities.

On the political-governmental front, cities face the prospect of diminishing influence as suburbs continue to become a larger and larger proportion of the population. Suburbs have long sought to isolate themselves from inner-city problems and discourage the immigration of poor and minority residents. Yet suburbs themselves have become increasingly varied entities, and state requirements on such matters as equal spending on education open up the possibility for cross-jurisdictional and intergovernmental alliances that bridge traditional local boundaries. Coalition building is a long recognized leadership activity, and city-suburb-state alliances provide a testing ground for this skill.

The social isolation of inner-city ghettos is perhaps the most formidable of the obstacles to be overcome, but it nevertheless must be confronted. The isolation is as much psychological as organizational, but it can be overcome only as people are enabled to act collectively and form their own coalitions in pursuit of their mutual aim to be incorporated into the mainstream of the nation's economic and civic life (O'Brien, 1975). Organizations of public-housing tenants and the continuing presence of inner-city churches indicate that collective activities are possible.

Changing the policy agenda thus requires that the conditions of regime building be addressed. Policies such as those to enhance human capital or expand affordable housing can be sustained and made effective only by assembling an appropriate regime with sufficient resources. The challenge for urban leadership is one of how to take existing conditions and modify them through a set of concrete steps that overcome the tendencies inherent in an unequal distribution of resources and motivation.

Redevelopment regimes provide evidence that arrangements can be devised to support complicated policy efforts. Opportunity-expansion regimes are a bigger challenge, but politics is an activity in which human beings take charge of their affairs and attempt to respond to the problems surrounding them. Regimes are political artifacts. And if they are to do more than reinforce society's inequalities, then political leadership must seek to build regimes that alter those inequalities. That presumably is the ideal that guides a democratic way of life.

REFERENCES

Bennett, L. (1989). Postwar redevelopment in Chicago: The declining politics of party and the rise of neighborhood politics. In G. D. Squires (Ed.), *Unequal partnerships: The political economy of urban redevelopment in postwar America*. New Brunswick, NJ: Rutgers University Press.

Bernick, M. (1987). *Urban illusions*. New York: Praeger.

Caro, R. A. (1974). *The power broker*. New York: Knopf.

Clavel, P. (1986). *The progressive city: Planning and participation, 1969-1984*. New Brunswick, NJ: Rutgers University Press.

Cohen, M. D., & March, J. G. (1986). *Leadership and ambiguity* (2nd ed.). Boston: Harvard Business School Press.

Darden, J., Hill, R. C., Thomas, J., & Thomas, R. (1987). *Detroit: Race and uneven development*. Philadelphia: Temple University Press.

DeLeon, R. E. (1990, August). *The triumph of urban populism in San Francisco?* Paper presented at the annual meeting of the American Political Science Association, San Francisco.

Dreier, P. (1989). Economic growth and economic justice in Boston. In G. D. Squires (Ed.), *Unequal partnerships: The political economy of urban redevelopment in postwar America.* New Brunswick, NJ: Rutgers University Press.

Fainstein, S., Fainstein, N. (1989). New York City: The Manhattan business district, 1945-1988. In G. D. Squires (Ed.), *Unequal partnerships: The political economy of urban redevelopment in postwar America.* New Brunswick, NJ: Rutgers University Press.

Ferman, B. (1985). *Governing the ungovernable city.* Philadelphia: Temple University Press.

Good, D. L. (1989). *Orvie: The dictator of Dearborn.* Detroit: Wayne State University Press.

Henig, J. R. (1982). *Neighborhood mobilization.* New Brunswick, NJ: Rutgers University Press.

Hochschild, J. (1989, August). *The politics of the estranged poor.* Paper presented at the annual meeting of the American Political Science Association, Atlanta.

Jones, B., & Bachelor, L. (1986). *The sustaining hand.* Lawrence: University Press of Kansas.

Kasarda, J. (1985). Urban change and minority opportunities. In P. Peterson (Ed.), *The new urban reality.* Washington, DC: Brookings Institution.

Kingdon, J. (1984). *Agendas, alternatives and public policies.* Boston: Little, Brown.

Krasner, S. D. (1983). *International regimes.* Ithaca, NY: Cornell University Press.

Levine, C. H. (1974). *Racial conflict and the American mayor.* Lexington, MA: Lexington.

Levine, M. V. (1989). The politics of partnership: Urban redevelopment since 1945. In G. D. Squires (Ed.), *Unequal partnerships: The political economy of urban redevelopment in postwar America.* New Brunswick, NJ: Rutgers University Press.

Logan, J., & Molotch, H. (1987). *Urban fortunes: The political economy of place.* Berkeley: University of California Press.

Lubove, R. (1969). *Twentieth-century Pittsburgh.* New York: John Wiley.

Lupsha, P. A. (1987). Structural change and innovation: Elites and Albuquerque politics in the 1980s. In C. N. Stone & H. Sanders (Eds.), *The politics of urban development.* Lawrence: University Press of Kansas.

Machiavelli, N. (1985). *The prince* (H. C. Mansfield, Trans.) Chicago: University of Chicago Press.

Mollenkopf, J. (1983). *The contested city.* Princeton, NJ: Princeton University Press.

Nelson, W. (1987). Cleveland: The evolution and decline of black political power. In M. Preston et al. (Eds.), *The new black politics: The search for political power.* New York: Longman.

O'Brien, D. J. (1975). *Neighborhood organization and interest-group processes.* Princeton, NJ: Princeton University Press.

Perrow, C. (1986). *Complex organizations.* New York: Random House.

Pressman, J. L., and Wildavsky, A. (1984). *Implementation.* Berkeley: University of California Press.

Preston, M. B. (1987). The election of Harold Washington. In M. B. Preston et al. (Eds.), *The new black politics: The search for political power.* New York: Longman.

Reed, A. (1988). The black urban regime: Structural origins and constraints. In M. P. Smith (Ed.), *Power, community, and the city: Comparative urban and community research,* Vol. 1. New Brunswick, NJ: Transaction.

Rich, W. (1989). *Coleman Young and Detroit politics.* Detroit: Wayne State University Press.

Rosdil, D. (1989, August). *Cultural and economic origins of progressive reform: The case of Burlington, Vermont.* Paper presented at the annual meeting of the American Political Science Association, Atlanta.

Sanders, H. T. (1987). The politics of development in middle-sized cities: From New Haven to Kalamazoo. In C. N. Stone, & H. Sanders (Eds.), *The politics of urban development.* Lawrence: University Press of Kansas.

Seyd, P. (1990). Radical Sheffield: From socialism to entrepreneurialism. *Political Studies, 38,* *335-344.*

Smith, M. P. (1988). *City, state, and market.* New York: Basil Blackwell.

Stone, C. N. (1980). Systemic power in community decision making. *American Political Science Review, 74,* 978-990.

Stone, C. N. (1989). *Regime politics: Governing Atlanta, 1946-1988.* Lawrence: University Press of Kansas.

Swanstrom, T. (1985). *The crisis of growth politics.* Philadelphia: Temple University Press.

Wilson, W. J. (1987). *The truly disadvantaged: The inner city, the underclass, and public policy.* Chicago: University of Chicago Press.

11

A Walk Around the Town:
Closing Observations
With Apologies to F. Engels

M. GOTTDIENER

IT WAS RAINING and as we got on the 67th street cross-town bus the crowd squeezed together more than usual. It was also 5:00 and everyone was eager to get home after a hard day's work. Rush hour and rain—not a good combination. The Puerto Rican kid in front of me bounced up the bus steps and then stopped in front of the fare box. He wore Walkman earplugs and apparently had been too busy listening to rap music to prepare his fare. Caught without a token he stood fumbling for the exact change. I pushed past and dropped my token in the box. I heard him yell as I passed by, "You're supposed to say excuse me!" I just ignored him and took a seat.

After paying he came down the aisle and stood in front of me. "You better watch yourself. You're supposed to say excuse me. That's the Right Thing to do." He then sat down at the back of the bus.

I noticed all the other passengers looking away. I was on my own. I thought about ignoring him. I decided not to. I turned toward him and asked "What happened?" He didn't answer. I pressed him on it, "I want to know, what happened?" Just then I noticed him fidgeting in his seat. Was he getting uncomfortable? And then, at once, he looked away. I had him! By looking away, he had backed down. "Fuck you," I said. It was over.

I was lucky. The chances were high that the kid carried a gun or at least a knife. Maybe he didn't and he backed down. It also wasn't his turf. But lately, people have told me, there isn't any neutral territory in the city anymore. The kids roam the city freely and attack when they can. They keep order, too. Adults are so terrorized by random violence, much of it perpetrated by youth gangs, that they live in fear of just such an incident. Everyone defers to them now, even the police. I was lucky.

*   *   *

I saw the two men every day, hanging out on the steps of the Presbyterian church on Park Avenue. It was their sanctuary. They were homeless but also probably insane. Both were black and at least one was elderly. He sat mournfully on the steps peering out over his ripped pants at the parade of well-heeled New Yorkers passing by. The younger was middle aged. He never looked up. Often he would pace back and forth on the sidewalk talking to himself. I passed them twice a day to and from work. I never saw them inside the church. Several times I was able to see how they slept—on the side of the building on the cement with their few belongings pressed in tight and newspaper all around acting as a coverlet. Some days it seemed too cold for them to live. I remember one day in particular. I was walking briskly to work. It was a busy rush-hour morning. Park Avenue was jammed with cars. The younger one was standing out in the street. I saw him out of the corner of my eye. I turned to look. He was crouched low and defecating in a big pile. Just like some Great Dane dog. Right on Park Avenue.

*   *   *

On the bus going to work one morning I took one of the few seats left. At Central Park West an old lady got on helped by a middle-aged white man. He supported her while paying the fare. She could barely walk. As they came down the aisle I rose and gave her my seat. "Here's a seat, Mom," the man said. "Thank you," the man said to me, "thank you very much." I thought then of Los Angeles, where I used to live and about how my life there was spent driving everywhere in my car. Road courtesy was rewarded with a wave and a pleasant smile. On the bus in New York I received something much more—public acknowledgment for a good deed.

There it was. Everyday small possibilities for doing good works. And being thanked publicly for it, too!

*   *   *

Some of the black students in my Introduction to Sociology class were complaining. All the material was "Eurocentric," they said. I was teaching from a white person's point of view. "What about the book chapter on

Racism," I asked? They invited me downstairs to see a video tape of Farakhan giving a speech. I went. There was a large crowd of students. He was raving about the Jews.

*   *   *

I needed a new belt. I decided to go to Bloomingdales even though it was getting close to the Christmas rush. I made the right choice, at least, for belt buying. This place wasn't suburban Los Angeles and I wasn't in a Mall. The department store had an entire room devoted to men's belts. I was in and out with my purchase quickly. Outside, the crowds of shoppers had gathered on Third Avenue. I noticed a group of men who were begging. They were all white and young, too; perhaps in their thirties. They were working the shoppers as they exited Bloomingdales. "Please, please, please," one said almost crying, "Just some money, some change, anything you can spare. I'm so hungry!" Another shouted at the backs of some fleeing shoppers. "You think I'm a bum. I'm a Viet Nam veteran. I can't get a job." They were dressed in rags. One was standing barefoot on the cold cement of the sidewalk.

Things like that happened to me in D.C. and I thought it most ironic. I mean, D.C. is the nation's capital. But in New York it takes on a grand scale. On one side, the glittering pleasure palaces of consumption up and down the midtown streets, like Trump's bronzed towers or the new Sachs Fifth Avenue store "Exclusively for Men." On the other side, the most abject scenes of poverty and desperation north of Mexico City or Rio. And just like the people's representatives in Congress, everyone seems to be adjusting to life lived with grim contrasts and to a society that has swallowed the suicidal pill of minimalist social policy.

*   *   *

I decided after several months of being back in New York City to visit my old neighborhood. I hadn't been there in over 30 years. I grew up in the South Bronx. I lived off of Southern Boulevard. I didn't really want to go back, even just to look. But something drew me there. Borrowing my friend's car, I decided to drive through.

I took the southern approach up through Bruckner Boulevard and exiting into the local streets. I passed the usual South Bronx landscape. There's nothing particularly "southern" about this experience. I drove by block after block of rubble; an abandoned house stood out in the middle of leveled buildings reduced to bricks and stone. Suddenly, I entered a block of intact houses. Many of the windows were broken. Yet, people also lived here. Children were playing in the street. There seemed to be a lot of garbage around. Was it ever collected? A stray, emaciated dog walked slowly into the intersection. I stopped the car and waited for it to pass. Then, as I drove on, the inhabited island had ended and there were more streets of rubble.

Memories were buried beneath all that stone. I remembered another time, long ago. The immigrant boys were standing outside Ludwig's candy store on Southern Boulevard. At that time I thought it was Ludwig's telephone number that I would see conveniently tattooed on his forearm. I was with my friends Jose, Israel, Lotzi, and Vinnie. We had just stolen some candy and glue from the store. It was easy. Sometimes we let Ludwig see us just so that he would run after us and scream on the street. We talked about getting the others in the gang and going over across the Boulevard to beat up the Irish. Somewhere lurking behind us all the city schools were waiting for us. They would perform their miracle and force us on our way to achievement. My dear friend Jose never made it. Too much Junk. Now even the buildings were gone.

About the Contributors

ROBERT D. BULLARD is an Associate Professor of Sociology at the University of California-Riverside. His Ph.D. degree was awarded from Iowa State University in 1976. He has held faculty appointments at the University of California-Berkeley, University of Tennessee-Knoxville, and Texas Southern University in Houston. His specialty areas include housing and residential patterns, urban land use, and environmental quality. He is the author of four books, the most recent is entitled, *Dumping in Dixie: Race, Class and Environmental Quality.*

YEN LE ESPIRITU, Assistant Professor of Ethnic Studies and Sociology at the University of California at San Diego, has written on ethnicity, immigration and intergroup relations. She is especially interested in the construction of ethnicity and panethnicity among Asian Americans and Latinos. Her graduate work at the University of California, Los Angeles, was supported by predoctoral fellowships from the American Sociological Association Minority Fellowship Program and the UCLA Institute of American Cultures. She has also been awarded a postdoctoral fellowship by the University of California President's Office to study panethnic relations among the major ethnic and racial groups in the United States.

JOE R. FEAGIN holds the Graduate Research Professorship in Sociology at the University of Florida. His research interests lie in urban sociology and racial and ethnic relations. He is author of the best-selling textbook, *Racial and Ethnic Relations* (1989) and of the classical study of discrimination, *Discrimination American Style* (1986). He is currently working on a book on the discrimination faced by middle-class black Americans.

RAY FORREST is a Senior Research Fellow at the School for Advanced Urban Studies, University of Bristol, England. He has written extensively on aspects of housing and social change. He is co-author, with A. Murie, of *Selling the Welfare State* (1988) and with A. Murie and P. Williams, of *Home Ownership: Differentiation and Fragmentation* (1990). Recent funded research has included work on the privatization of public housing, the changing sociology of home ownership, and the position of marginal groups in urban areas. He is currently involved in a major study of housing and labor mobility.

M. GOTTDIENER is currently Professor of Sociology at Hunter College of the City University of New York, and director of the City As Lab project. He is the author of six books including *The Social Production of Urban Space* (1986), *The Decline of Urban Politics* (1987), and, with N. Komninos, *Capitalist Development and Crisis Theory* (1989). He works in the areas of urban theory and policy, political economy, semeiotics, and cultural studies. He is currently completing a text, *The New Urban Sociology,*.

DAVID IMBROSCIO is a doctoral candidate in the Department of Government and Politics at the University of Maryland. His current research focuses on urban regime formation and alternative strategies for local economic development.

GEORGE KEPHART is an Assistant Professor of Sociology and a Research Associate at the Population Issues Research Center at Pennsylvania State University. His research interests include migration, population redistribution, urban policy, and the labor force.

IVAN LIGHT is Professor of Sociology at the University of California, Los Angeles, and currently Director of the Interdisciplinary Immigration Research Project. He is the author of three books: *The Ethnic Enterprise in America, Cities in World Perspective,* and, with Edna Bonacich, *Immigrant Entrepreneurs.*

LINDA McDOWELL is a Senior Lecturer in Geography/Urban Studies at the Open University, England. Her interests are in feminist theory, gender divisions of labor, and the impact of economic restructuring on women's waged and unwaged work. She has written articles in this area as well as on housing policy and is author of *Landlords and Property* (1989) with John Allen, and of *The Transformation of Britain* (1989) with Michael Ball and Fred Gray.

MARION E. ORR is a doctoral candidate in the Department of Government and Politics at the University of Maryland, College Park. His research interests are in urban politics and policy and black politics.

CHRIS G. PICKVANCE is Professor of Urban Studies and Director of the Urban and Regional Studies Unit at the University of Kent at Canterbury. His research interests are in local government, housing, urban protest, urban and regional policy, and in comparative analysis and locality studies. He is co-editor, with M. Harloe and J. Urry, of *Place, Policy and Politics: Do Localities Matter?* (1990), and, with E. Preteceille, of *State Restructuring and Local Power: A Comparative Perspective* (1990). He is a founder-member of the Editorial Board of the *International Journal of Urban and Regional Research,* and President of the ISA Research Committee on the Sociology of Urban and Regional Development.

BRYAN R. ROBERTS is Professor and C. B. Smith, Sr., Chair in U.S.-Mexico Relations at The University of Texas at Austin. Before coming to Texas in 1986, he spent 22 years in the Department of Sociology of the University of Manchester, specializing in comparative social structure, development, and urbanization, carrying out field research in Guatemala, Peru, Mexico, and Spain. His publications include *Organizing Strangers* (1974), *Cities of Peasants* (1978), and with Norman Long, *Miners, Peasants and Entrepreneurs* (1985), "Urbanization, Migration and Development," *Sociological Forum,* 4,4(1989), and "Peasants and Proletarians." *Annual Review of Sociology,* 16:353-77 (1990).

GREGORY D. SQUIRES is Professor of Sociology and a member of the Urban Studies Program Faculty at the University of Wisconsin at Milwaukee. He served from 1977 to 1984 as a research analyst with the U.S. Commission of Civil Rights. His research has focused on the racial effects of uneven urban development during the post-World War II years. Recent publications include his co-authored book *Chicago: Race, Class and the Response to Urban Decline* (1987) and his edited book *Unequal Partnerships: The Political Economy of Urban Redevelopment in Post-War America* (1989).

CLARENCE N. STONE is Professor of Government and Politics at the University of Maryland. His most recent book is *Regime Politics: Governing Atlanta, 1946-1988*. His current research interests are power, leadership, and agenda-setting.

RALPH B. TAYLOR is a Professor in the Department of Criminal Justice and Associate Dean for Graduate Studies, Research and External Funding in the College of Arts and Sciences at Temple University. He has held positions at Virginia Tech and Johns Hopkins University. His research interests in the last few years have centered on the ecology and environmental psychology of crime and fear of crime. He is the editor of *Urban Neighborhoods* (1986) and the author of *Human Territorial Functioning* (1988).